The SecRet Life of the BRAIN

The Secret Life of the BRAIN

Unlocking the mysteries of the mind

Alfred David

FIREFLY BOOKS

A FIREFLY BOOK

Published by Firefly Books Ltd. 2019
Copyright © 2019 Octopus Publishing Group Ltd
Text copyright © 2019 Alfred David

First printing

ISBN-13: 978-0-2281-0176-5

Library of Congress Control Number: 2018963776

Library and Archives Canada Cataloguing in Publication is
available from Library and Archives Canada

Published in the United States by
Firefly Books (U.S.) Inc.
P.O. Box 1338, Ellicott Station
Buffalo, New York 14205

Published in Canada by
Firefly Books Ltd.
50 Staples Avenue, Unit 1
Richmond Hill, Ontario L4B 0A7

Edited and designed by Tall Tree Limited

Printed and bound in China

Contents

Introduction

Through the centuries, anatomists and physicians have prodded, sliced and dissected their way through the brain, but its mysterious inner workings have proven difficult to uncover.

The brain is essential for life, responsible for our thoughts and for our feelings. It makes us angry, scared or in love, and it is responsible for moments of stunning human creativity. Mozart's brain produced spine-tingling music. The brains of Newton, Einstein and Hawking probed their understanding of the universe. And Darwin's brain paved the way for understanding why we have a brain in the first place. But while organs such as the heart have obvious moving parts, with obvious signs of function, the brain's structure, variously described as having the texture and consistency of soft tofu, a ripe avocado or (even) scrambled eggs, seems to belie its ingenuity. It contains chambers with seeping fluids, and is supported by a vigorous blood supply, meaning that it is exceptionally

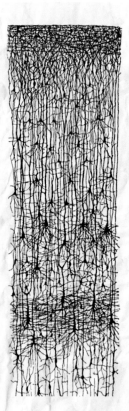

Spanish pathologist Santiago Ramón y Cajal (1852–1934) made hundreds of detailed drawings of the nerve networks he had discovered in the brain.

demanding of food and oxygen. But the brain seems to be singularly static and silent. That is, until we clamp electrodes to a living scalp and see all the electrical activity generated by the organ within.

Function From Failure

Since the time of the ancient Greeks, anatomists have dissected bodies and mapped the parts of the brain, from its flabby lobes to its central stem, and given them names. They even managed to work out what some of them did by doing grisly experiment on living animals, cutting into parts of the nervous system to see what effect it had. Later, on the battlefield and elsewhere, physicians studied the unfortunate victims of specific brain damage. By seeing how injuries or lesions caused by disease affected the way their patients behaved, physicians slowly uncovered the secrets of the brain parts discovered by the anatomists. Post-mortems on damaged brains help to reveal how different parts of the brain are involved in speech, personality or even moral awareness. But exactly how this is all controlled, and what the brain actually did in a physical sense, was still a secret. Muscles work by twitching, blood works by flowing. But what causes a thought inside the brain? The microscope brought some answers.

Secrets Under Magnification

Towards the end of the nineteenth century, biologists around the world were perfecting techniques that would help them understand how living things worked at the microscopic level. Their work showed them that all parts of bodies were made up of tiny living structures called cells. Indeed, many people were now convinced that everything the body did – how it moved and even how it thought – depended on cells. In Barcelona, a young pathologist called Santiago Ramón y Cajal had made his name by studying cholera and tissues before turning to the brain. Using a special staining technique, Ramón y Cajal revealed single brain cells among their mass of neighbours. In this way, he helped to prove that the brain followed the rest of the body's rules of organization: it, too, was made

up of communities of cells. This work laid the foundations of modern neuroscience. Today, biologists practically take it for granted that our bodies are made up of microscopic living cells: nowadays schoolchildren can peer down simple microscopes to look at cheek cells and onion cells. But not everyone in the past had been convinced. At the time of Ramón y Cajal, there was a competing theory as that the nervous system was made up of a single continuous network of tiny fibres. Ironically, in 1906 Ramón y Cajal shared a Nobel Prize for his work on the brain with an advocate of the network theory: Italian biologist Camillo Golgi dismissed the idea that the brain was made up of discrete cells. Golgi had, in fact, invented the very staining technique that Ramón y Cajal had used to make his discoveries. But in the end, there was no doubt that that brain – like any other organ – is made up of cells, and Ramón y Cajal's exquisite illustrations of branching brain cells are still shown to students today.

Brain Cells at Work

The demonstration that the brain is made of cells marked a seismic shift in neuroscience. It meant that the brain needed to be understood in terms of cellular things – like cytoplasm and membranes. As we shall see in the opening chapters of this book, this helps us appreciate how brains work, from the way they transmit the electrical impulses associated with thought processes and reactions to how they hold on to memories. In that respect, there is nothing mysterious about how the brain works: it all comes down to molecules and electrical charges, the stuff of any living matter. But the enormous organizational complexity of the billions of cells in a human brain means that understanding it is not easy. Brain cells are not randomly arranged: they are linked to one another in very particular ways. Their long, spindly fibres can connect to others over long distances through cable-like highways called tracts, while different regions of the brain are specialized in controlling vital functions, burdening our thought with emotion, and so on. This spatial organization is explained in the middle section of this book.

In social gatherings, our complex brains are constantly monitoring the subtle cues that we receive from others. Our brains allow us to read the emotions of others, to share their joy and to feel their pain.

Beyond Brain Cells

Brain cells are the vital components in the brain's electrical circuitry, but that is only part of the story when it comes to the secret life of the living brain. Brains do not work in isolation from the body's surroundings any more than they work isolated from other organs. Our brains react and respond to outside cues, some of which, in the complex world of our highly social species, come from other brains that are ticking over in other bodies. As we will see in later parts of this book, this means that our brains can get tuned in to the brains of our family and our companions. We behave differently in social settings and our brains are affected by the experience of things like parenthood.

It is an extraordinary thought, one of many that we can enjoy with our remarkable brain, that this entire gamut of complexity arises from an organ that develops like any other, beginning with a microscopic cluster of cells in a tiny embryo. In the close of this book, we will trace the life of a human brain from conception through to the rarefied wisdom of old age. In the womb, the growth of our brain in many respects parallels a far more ancient process that produced complex nervous systems out of simpler ones, millions of years ago. Our brain is not just a product of embryonic development, it is also the culmination of a process of evolution that has made our species so remarkable in its accomplishments, both good and bad. There are more than ten times the number of brain cells in a single modern human head than there are human beings on the face of the planet. That, in itself, is certainly food for cerebral thought.

Chapter 1
WHY HAVE A BRAIN?

Coordinating Bodies

The brain is part of an animal's nervous system, which has the job of controlling the activities of the living body. Although the simplest animals alive today lack a true brain, their nervous systems can still coordinate behaviour.

We share our planet with trillions of living things. Some are so small that you need a microscope to see them, but every one of them is busy controlling the complicated business of staying alive. For some, such as plants and bacteria, control depends upon chemical mixtures that signal a particular response. But the bodies of animals contain special cells that fire thought processes and trigger muscles to twitch into action. Compared to plants, animals move quickly and think hard. And most of them think with their heads: they have a brain.

In complex animals, including humans, the brain is the ultimate organ of control. A hard-working human brain looks as unremarkable as a resting one, but each brain is capable, silently and secretly, of feats that are more impressive than those of the most powerful computer. Indeed, its microscopic arrangement, with billions of interconnected nerve cells, has been described as the most complicated working thing in the entire known universe. But brains are expensive to maintain, and use up a great deal of energy: in a human,

Even animals that do not have a brain, such as jellyfishes, have some electrical circuitry inside their bodies – a nervous system – that helps them to react to their surroundings.

Why Have a Brain?

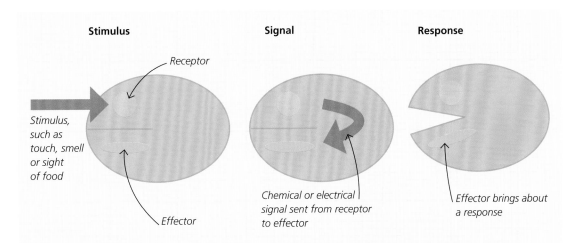

Stimulus

Signal

Response

Receptor

Stimulus, such as touch, smell or sight of food

Effector

Chemical or electrical signal sent from receptor to effector

Effector brings about a response

The simplest control system in an organism provokes an automatic response when a receptor (sensor) is stimulated. It involves a chemical or electrical signal being sent from the receptor to the part of the body that responds. This effector could be a gland or a muscle.

the brain uses about a fifth of the body's total energy budget. And the fact that the simplest animals manage their bioelectric circuitry without any real brain begs the question: why have a brain at all? To answer that question, we need to begin by looking at an arrangement inside the body that is home for any working brain: the nervous system.

How to Control a Body

What do we mean by "control"? Life processes, such as using food, breathing and growing, involve countless chemical reactions that occur simultaneously, and a fine balance is needed to get them to work. On top of everything, the living body is exposed to constant changes, and it has to react to these changes to keep working properly. For a beating heart, this could mean speeding up when the body starts exercise, or just pumping at the right pace when at rest. For a hungry animal, it means moving towards food, but also not straying too close to danger. In countless ways, the body has to control what it is doing.

All living things sense their surroundings in one way or another. They have special sensory cells that can detect stimuli: changes in their surroundings that can trigger a bodily response. A chain reaction in the body ensures that each stimulus leads to an appropriate response. The simplest chain might involve the sensory cells releasing a chemical that seeps through the body, and this chemical causing other body parts to respond. This is what happens in plants when their shoots bend towards light. It happens all the time right inside our own bodies when hormones flood through our blood, such as when a meal prompts the body to deal with the sudden surge of sugar. But a faster communication between sensor and responder is possible when it involves the flow of a bioelectric impulse. Nervous systems, found uniquely in animals, including us, do just that. And they explain why animals are usually quicker at responding to stimuli than plants. Over large distances, bioelectric impulses are much faster than seeping chemicals.

A nervous system is made up of special cells that can carry electrical signals. They provide the link between sensory cells and cells called effectors that can bring about the response. A brain is part of this nervous system. Some effectors are glands that, upon being triggered, give off chemicals that themselves act as more triggers. But other effectors have a more dramatic, observable effect: these are the twitching muscles.

Most muscles can only ever contract (shorten) when stimulated by the nervous system, before returning to their relaxed (longer) state. This means that the nervous system must coordinate different muscles to move a part of the body one way and then back again. The simple act of a human walking upright on two legs involves the nervous system collecting information from many different sensors and coordinating many muscle groups.

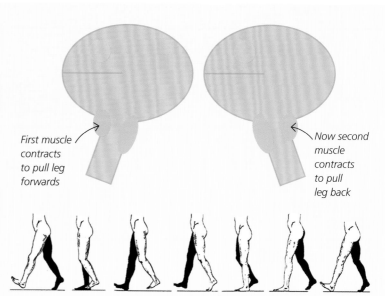

First muscle contracts to pull leg forwards

Now second muscle contracts to pull leg back

Twitching Muscles

Among all living things, only animals can consistently and quickly move in the way they do. This movement is possible because an animal's nervous system is linked to muscles. Each muscle is made up of long, tapering cells that are stacked together and, like the nerve cells, can carry bioelectric signals. But when the nervous system triggers a muscle, it moves. More precisely, it shortens in a twitching motion called contraction. By controlling muscles in different parts of the body, an animal's nervous system can control all its movement in a coordinated way. For instance, in a walking animal, bioelectric triggers from the nervous system ensure that different muscles of the leg contract in just the right sequence so that it bends and straightens, and the animal ends up putting one foot in front of the other.

Nervous Systems Without Brains

The simplest kind of nervous system imaginable would involve a single nerve cell between a sensor and a muscle. Such a system would be very good at producing a rapid-firing response: whenever the sensor was stimulated, the muscle would twitch. But the system would be automatic and would

lack any flexibility. The animal would have no choice in the matter of whether the muscle should twitch or not.

In reality, even the simplest nervous systems are made up of many nerve cells that interconnect to form a network. Nerve cells, called neurons, can do this because each one is armed with spindly fibres that radiate outwards from the main cell body, carrying their electrical signals with them. The fibres of adjacent neurons appear to touch and interconnect, but as we will see in the next chapter, there are tiny gaps where they meet. A network of neurons like this immediately offers more possibilities for controlling behaviour. There is now no longer a single pathway between one sensor and one muscle. Instead, the signal from one sensor has the option of travelling through a number of possible routes to make different muscles respond. Our brains contain complicated networks of densely packed connections, but the principle works even in the simplest animals without a brain, albeit on a smaller scale.

Although they are brainless, jellyfishes, sea anemones and corals have a network of neurons. It runs through the flesh of the animal to control parts such as moving tentacles. And although it is nowhere near as densely packed as the neurons of our brain, this "nerve net" is clearly adequate for their purposes. When tiny planktonic animals drift within reach, sensors are triggered and the nerve net springs into action to coordinate a set of muscles that makes the tentacles squirm towards their prey. In a swimming jellyfish, a more regular coordinated pattern of action helps cause the pulsating movements that drives the animal through the water.

Forward-Thinking

The reason jellyfishes and anemones do not have a brain might seem obvious: they don't even have a head. Each animal consists of a "stem" with a single gut opening at one end that functions as both a mouth and an anus, surrounded by a ring of tentacles. Jellyfishes swim through water, with their mouth-anus facing downwards, while anemones sit on the seabed with theirs facing up. But neither has a head. In fact, they don't really have a front end and back end. Their bodies are radially symmetrical, and they have a sense of "up" and "down", but not "forwards" and "backwards".

For them, a spreading nerve net works well because it helps the animal pick out stimulation coming from all directions at once.

Half a billion years ago, when animals were first evolving, those like the jellyfishes and anemones, adopted radial symmetry as their way of life and stuck to it. But some worm-like creatures evolved to move forwards instead. They became bilaterally symmetrical, meaning that the body could be divided down the middle into two halves that were mirror images of one another. One end became the front (with the mouth) and the other became the back (with the anus). When moving forwards, it was better for them to concentrate many of their sensors at the front end – the end that would be facing any new stimuli. And with a bigger frontal battery of sensors, the nervous system at the front swelled to process this incoming information. A head evolved to accommodate the swelling, which became the brain. Today, most animals, including humans, have a forward-facing head packed with sense organs and a brain. Ultimately, a brain is the legacy of our ancestors choosing to move forwards through the world.

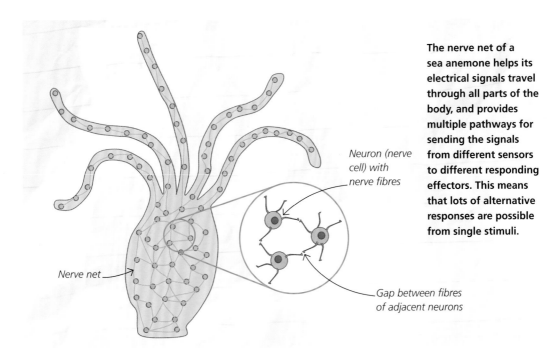

The nerve net of a sea anemone helps its electrical signals travel through all parts of the body, and provides multiple pathways for sending the signals from different sensors to different responding effectors. This means that lots of alternative responses are possible from single stimuli.

Neuron (nerve cell) with nerve fibres

Nerve net

Gap between fibres of adjacent neurons

A Brain in the Driving Seat

In forward-moving animals, the brain inside the head is the control centre of the entire nervous system, while nerve cells communicate between the brain, sensors and muscles.

Although practically every part of our body has sensors and nerve cells, our head is the primary place concerned with processing the information that we gather from the world around us. The evolution of a head in animals was, literally, a forward-facing innovation. There is even a term for it: cephalization. Animals with brain-packed heads are the best at moving forwards: crawling on the ocean floor, swimming in open water, or moving over land or in the air. It is far better to have a control centre at the front of the body, which is the first part to encounter information from newly explored places. Cephalization was a breakthrough in the evolution of animal life.

Although the head is not discretely separate in many animals (in the sense that they have no neck), practically all animals alive today that are more complex than a jellyfish or anemone have some sort of brain.

What is a Brain?

As forward-facing animals evolved to become bigger and more complex, their nervous systems became more complex, too, and the part of the

Animals moving forwards in search of food have their sensors packed in their front end, which will be the first part of the body to come across a potential meal.

Why Have a Brain?

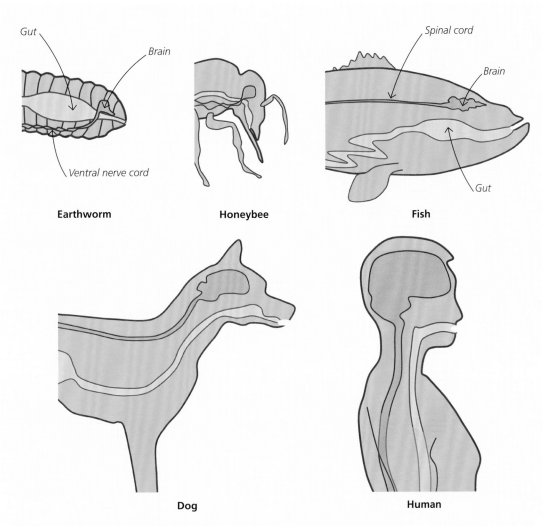

Gut

Brain

Ventral nerve cord

Earthworm

Honeybee

Spinal cord

Brain

Gut

Fish

Dog

Human

The brains of animals vary a great deal in size and complexity. Invertebrates (animals without a backbone, such as earthworms and honeybees) have smaller brains in proportion to the size of their body than most vertebrates (animals with a backbone, such as fishes, dogs and humans).

nervous system at the front end, bearing the brunt of the sensory barrage, became more complex than other parts: it became the brain. In the simplest sense, a brain is a dense mass of nerve cells that controls and coordinates behaviour: it receives signals from sensors around the body and sends out other signals to parts of the body, such as muscles, that can respond. It can also store information, in the form of memories, which can affect what an animal does.

Animals without a backbone, known as invertebrates, include animals such as worms, insects and snails. The brains of most invertebrates are little more than a small blob of nerve cells, although they still have considerable processing power. Some scientists prefer to use a different term for an invertebrate mini-brain: they call it a cerebral ganglion. However, in this book, we shall define any central coordinating "blob" in the head of an animal as a brain.

A Centralized Nervous System

With its front-facing brain, the nervous system of a worm and other front–back animals differs from that of a radial jellyfish is another important way: it has become centralized. Instead of a mesh of nerve cells spreading through the body, nerve cells are concentrated not only into a brain, but also in its conduits. Running down the length of the body, from brain to rear, are solid cords of nerve cells. The simplest worms have two or more cords, but in more complex animals there is a single cord. The brain and cord (or cords) make up the animal's central nervous system (CNS). All along the CNS, finer branches, called nerves, extend out from the brain and cord to penetrate, or innervate, the rest of the body. These nerves carry signals from the

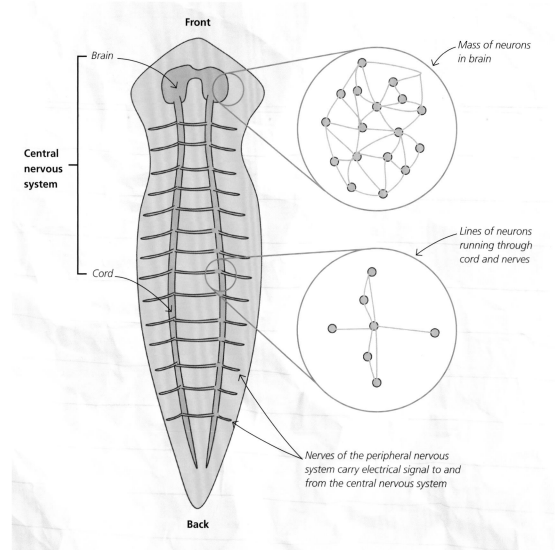

Front

Brain

Central nervous system

Cord

Mass of neurons in brain

Lines of neurons running through cord and nerves

Nerves of the peripheral nervous system carry electrical signal to and from the central nervous system

Back

The nervous system of animals with front and back ends, such as this flatworm, is centralized, with the brain in the head and one or more nerve cords running down the length of the body. Branches from this central nervous system are nerves of the peripheral nervous system.

Why Have a Brain?

Problem-solving, creativity and emotional experiences are all made possible by a large brain packed with billions of neurons that are interconnected in complicated ways.

body's sensors into the CNS, as well as other signals coming away from the CNS towards the muscles. Altogether, the nerves make up the peripheral nervous system (PNS). The job of the CNS is to control and coordinate these signals as they pass from sensors to muscle. The brain contains more nerve cells with more complex interconnections, so it plays a bigger role in this process than the cords.

In invertebrates such as insects the nerve cord runs down through the underside of the animal, from its head, through the belly beneath its guts and other major soft organs, right to its rear end. By contrast, in the bodies of vertebrates – animals with a backbone (fishes, amphibians, reptiles, birds and mammals) – the single cord runs down the back. The entire CNS of a vertebrate animal is protected within parts of the hard skeleton. The brain is encased in the skull, while the cord is encased in the spine. This is why the cord of backboned animals is properly called the "spinal cord".

What Can Complex Brains Do?

In the most complex animal brains, including ours, there is far more going on than automatically coordinating responses to incoming information. Certainly, the most vital processes of life, such as heartbeat or breathing, are rigidly controlled in this way, but brains have "higher" centres too that make our behaviour much more complicated than that. Bigger brains with more neurons are better at storing memories, while their processing powers help with decision-making so that responses are not always fixed and automatic. In short, brains help animals solve problems, such as "where is the best place to find food in this forest?" or "what is the answer to this exam question?" The most complex brains of all – the brains inside the heads of humans and our closest living relatives – also produce sensations such as pleasure and fear. This means that the brainiest creatures experience emotions and even have individual personalities. How the brain does all these things is a big part of this book.

After collecting nectar, bees return to their hives and tell other bees where the flowers are by performing a "waggle dance". The dance communicates the direction and distance to the flowers. These are complex behaviours, but it appears that the bees have little choice but to perform them.

Can Creepy-Crawlies Think?

Before we can understand our own mental abilities and those of other animals, we need to ask: what, exactly, do we mean by thinking? At least as far as humans are concerned, thinking involves processes in the brain, and these processes consist of bioelectric signals passing along nerve cells. Although there is no consensus that goes any further than that, thinking is generally considered to result in ideas, realizations or (when solving a problem) conclusions. The brains of invertebrates, such as worms and insects, undoubtedly play host to bioelectric signals and can even store memories, but their simple structures place limitations on any processing that is more complex than that.

The biggest limitation may be in the flexibility of their responses. If thought processes encompass any mental processing, then certainly a bug can think. But much of its behaviour involves automatic responses and routines. A honeybee, for instance, is "programmed" to collect nectar from flowers and to communicate the location of the flowers to other members of the hive in a very precise set of moves. The bee needs to learn where the food is before it can communicate its location. This involves a degree of flexibility and the need to make choices. But a bee lacks the capacity to make decisions that override its "programming", so it would not, for instance, decide on a whim to take a day off from its work!

Why Have a Brain?

Chapter 2
HOW NERVE CELLS WORK

Nerve Cells

The nervous system is made up of microscopic nerve cells called neurons. We can only properly understand how the nervous system works by taking a closer look at the workings of these individual cells.

Practically every part of a living body is made up of microscopic cells, and the brain is packed with more than most: there are about 100 billion of them in an adult human brain. Its neurons work by passing on signals, but that is only part of the story. The incredible complexity of a brain, allowing it to react to stimuli, store memories, make us feel emotions and give us a sense of "self", ultimately comes down to the bioelectric activity of its cells and how they communicate with one another.

All animal cells contain the same kinds of components. Each one is filled with a runny jelly-like cytoplasm that is bound on the outside by a very thin layer called the cell membrane.

Inside the cytoplasm, most kinds of cell have a nucleus, which contains DNA, the cells' genetic material. Inherited through generations, DNA controls what cells and bodies do and how they develop. But the fate of every cell inside the body is also determined by the conditions it is exposed to in its surroundings. In this way, cells become specialized in what they do as the body develops. The brain cells of a human are human because of their DNA, and they control distinctly human-like behaviour. But they are also bathed in chemicals that can affect how you feel. By looking at the shape, structure and workings of these cells, we can begin to understand the capabilities of the brain.

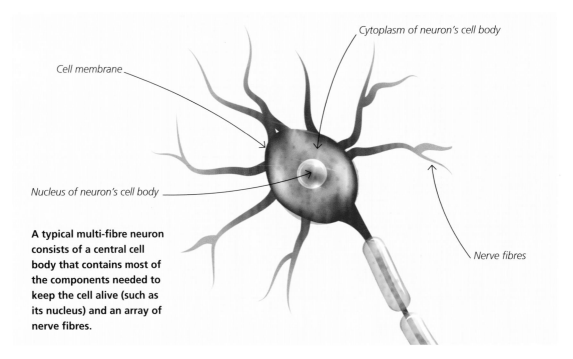

Cytoplasm of neuron's cell body

Cell membrane

Nucleus of neuron's cell body

Nerve fibres

A typical multi-fibre neuron consists of a central cell body that contains most of the components needed to keep the cell alive (such as its nucleus) and an array of nerve fibres.

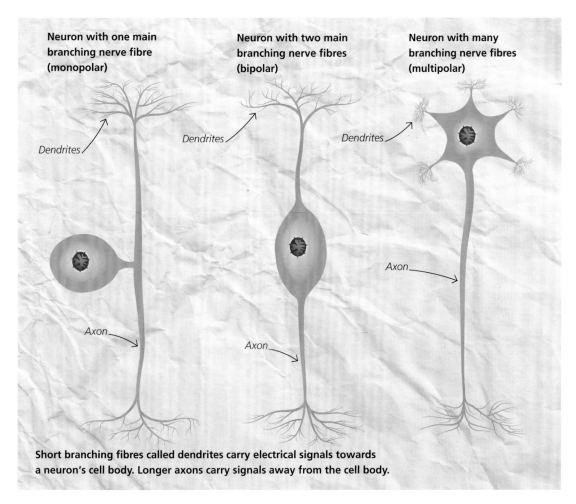

Neuron with one main branching nerve fibre (monopolar)

Neuron with two main branching nerve fibres (bipolar)

Neuron with many branching nerve fibres (multipolar)

Dendrites

Dendrites

Dendrites

Axon

Axon

Axon

Short branching fibres called dendrites carry electrical signals towards a neuron's cell body. Longer axons carry signals away from the cell body.

Parts of a Neuron

The most prominent feature of a neuron (nerve cell) is its arrangement of long, thin fibres. A neuron looks as though someone has taken an ordinary-looking cell and pulled points of its cytoplasm out into very long spindly threads. As a result, a neuron consists of a cell body that contains the nucleus and other essential structures, and an attached arrangement of thread-like nerve fibres.

Each thread is still bound by the oily cell membrane (something that is especially important, as we shall see), so the oily surface of a typical neuron is very big indeed. The fibres of some kinds of neurons can be a phenomenal length: the fibres of neurons running through our limbs can be more than 1 metre (3 ft) long.

Neurons are easily the longest kinds of cell in existence. The record-breaking fibres of a neuron serve an obvious purpose: they carry the bioelectric signals of the nervous system over long distances uninterrupted. This helps to explain why the nervous system can react so quickly to stimuli. But the arrangement of fibres also varies among neurons. Some have a single extra-long fibre with a cell body at one end. Others have two long fibres, with the cell body in the middle, or have a multitude of fibres. Nerve cells in the brain generally have the most fibres of all. Remarkably, each cell can have hundreds or even thousands of fibres radiating from a central cell body, ready to communicate with their neighbours.

Types of Neuron

Neurons are grouped according to their role in the transmission of signals through the nervous system. Some specialize in collecting signals from sensors and sending them into the central nervous system: either the brain or spinal cord. These are called sensory neurons. Others work to send signals back out from the CNS to muscles or glands. These are called motor neurons. Sensory and motor neurons usually consist of just one or two fibres arranged in a line, and are best organized for long-distance transmission. Inside the body, these fibres are arranged as bundles, which make up the nerves of the body.

Inside the brain and spinal cord, there is also a third type of neuron, called an association neuron (also referred to as the relay neuron or interneuron). Association neurons are tasked with completing the circuit between the sensory and motor pathways. For some fast-acting reflex actions involving signals through the spinal cord, association neurons do little more than ferry the signals passing through. But association neurons in the brain do a lot more than this. Indeed, they are responsible for some of the most complex brain activities that work on the signals to give us memories, help us solve problems and experience emotions.

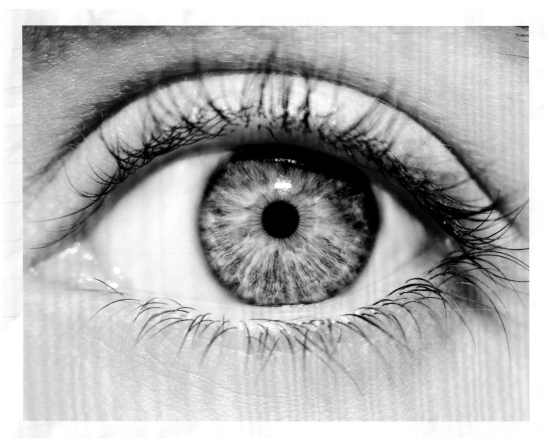

In bright light, your pupils contract in a reflex action. Sensory neurons send a signal from the eye to the brain, which fires motor neurons to contract certain muscles in the iris.

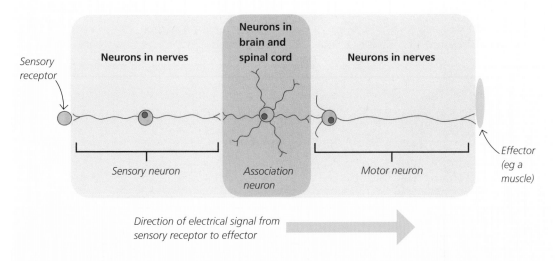

Sensory receptor

Neurons in nerves

Neurons in brain and spinal cord

Neurons in nerves

Sensory neuron

Association neuron

Motor neuron

Effector (eg a muscle)

Direction of electrical signal from sensory receptor to effector

**Sensory and motor neuron fibres occur in nerves of the peripheral nervous system.
Associations neurons are found in the brain and spinal cord: the central nervous system.**

Insulated Neurons

Outside the brain and spinal cord, most of the long-distance signalling along nerve fibres is aided by a special feature that helps to speed up the transmission. Each fibre is enclosed in a glistening coat of fatty material called the myelin sheath. Electrical charges cannot easily pass through fat, which means that the myelin sheath serves to insulate the fibres, just like the plastic insulation that protects electrical wiring. In this way, the sheath stops the electrical charges from escaping from the fibres, ensuring that they are properly funnelled through the nervous circuitry. As we shall see, it also makes the signals move even faster in their passage between the CNS and sensors or muscles.

These insulated fibres are also found in parts of the central nervous system, where dense concentrations of myelin make the tissue look white and glistening, giving it its name "white matter". The parts of the CNS that are not myelinated are, instead, packed more with cell bodies. The tissue here looks darker and is called "grey matter".

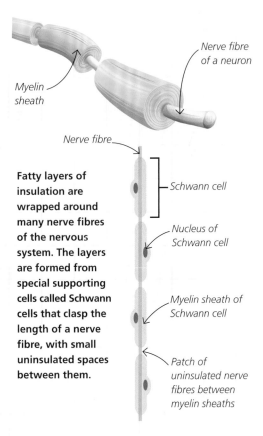

Nerve fibre of a neuron

Myelin sheath

Nerve fibre

Fatty layers of insulation are wrapped around many nerve fibres of the nervous system. The layers are formed from special supporting cells called Schwann cells that clasp the length of a nerve fibre, with small uninsulated spaces between them.

Schwann cell

Nucleus of Schwann cell

Myelin sheath of Schwann cell

Patch of uninsulated nerve fibres between myelin sheaths

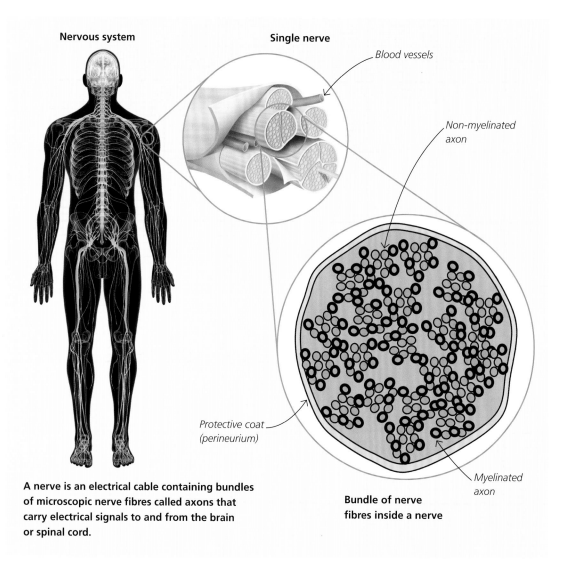

Nervous system

Single nerve

Blood vessels

Non-myelinated axon

Protective coat (perineurium)

Myelinated axon

A nerve is an electrical cable containing bundles of microscopic nerve fibres called axons that carry electrical signals to and from the brain or spinal cord.

Bundle of nerve fibres inside a nerve

How Nerve Fibres are Arranged into Nerves

The longest nerve fibres of a neuron usually carry electrical signals away from the cell body. These are called axons. Bundles of axons are bound together inside thicker cables, which are the nerves. Each bundle is wrapped inside a protective coating and several such bundles are packed together along with blood vessels to make a single nerve. The blood vessels provide food and oxygen for the working neurons. Generating the electrical activity necessary for a nerve fibre to carry bioelectric signals demands a lot of energy. This is provided by energy-rich food, such as sugar, which releases energy in chemical processes that use oxygen, called respiration.

The body contains a fixed arrangement of nerves that is set in place during early development. Some nerves carry only fibres of sensory neurons or motor neurons, but others have a mixture of both. Most of the main nerves, called spinal nerves, emerge and branch from the spinal cord. But others, called cranial nerves, such as the optic nerves that link to the eye, connect directly with the brain.

Bioelectric Signals

The signals of the nervous system are carried along the oily membranes of the neurons. These nerve impulses are generated by a stimulus, and may trigger new impulses in adjacent cells or make a muscle contract.

All living things need to respond to their surroundings and do so with the speed that is needed to stay alive. Plants and other "rooted" organisms, such as fungi and algae, generally respond quite slowly, but they rarely have a need to do it any more quickly. But animals need fast responses, not least because they eat other living things. This means that they have to move around to find their food, or even chase down other things that are their prey. Plants and algae, by contrast, bask in sunlight to make their food, while fungi take it from where they are growing.

No wonder that the most lightning-fast physical responses only happen in animals.

But lightning-fast responses need lightning-fast signals through cells. The simplest signal, a chemical message such as a hormone, won't do. It takes time for the chemical to seep or flow from one place to another. Chemical messages work for plants and fungi, and are useful in certain circumstances in animals (such as sexual development or triggering food-storage after a sugar rush), but are no good for fast moves. Instead, animals rely on something that is literally shocking. They use electrical charge.

Both chemical hormones and electrical impulses are used to signal around the body, but bioelectric impulses are much, much faster.

Electrically Charged Cells

Remarkably, all living things, from the simplest bacteria to plant and animals, use electric charge in their everyday lives. Living cells carry charges on their membranes, though the charges are very small. Cell membranes carry static electricity, just like the static electricity that can build up to stick paper to a balloon. Charges on particles come in negative and positive forms: opposite charges attract, while like charges repel. Paper sticks to a balloon because they carry opposite charges. In the same way, there is a small charge difference across the thin membrane of a typical cell. Usually, the outer surface has more positive charge, and the inner surface is negative in comparison. Cell membranes contain special kinds of protein molecules that keep the membrane charged up by, for instance, pumping out more positive charge than they bring inside. But given that cells expend energy doing so, what purpose does it serve?

Many organisms use the charge difference across their membranes as a potential store of energy, a bit like a battery. By letting opposite charges flow through at discrete points, they can use them to do work. But for fast-reacting animals, charged cell membranes also do something else: they carry signals. And these signals form the basis of the entire nervous system, including the brain.

The fastest movements in the animal kingdom, such as that of a chameleon's tongue, are possible because of the rapid-firing electrical signals of neurons and muscle cells. These same signals are responsible for the fast-acting thought processes in the brain.

How Nerve Cells Work

Nerve impulse

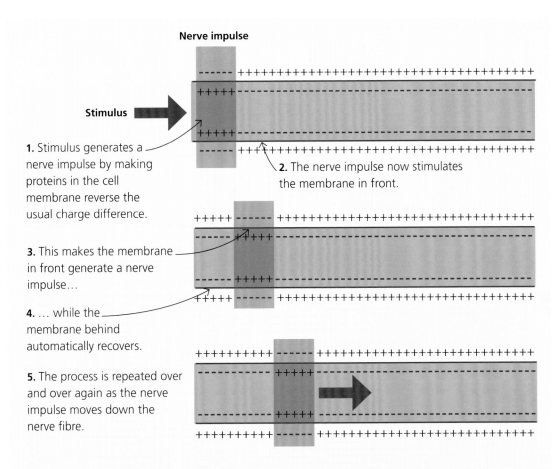

Stimulus

1. Stimulus generates a nerve impulse by making proteins in the cell membrane reverse the usual charge difference.

2. The nerve impulse now stimulates the membrane in front.

3. This makes the membrane in front generate a nerve impulse…

4. … while the membrane behind automatically recovers.

5. The process is repeated over and over again as the nerve impulse moves down the nerve fibre.

The long nerve fibres of neurons carry nerve impulses as fast-moving "waves" of electrical activity: a zone of excitation. A wave of charge reversal stimulates the membrane in front of the stimulus source, while the membrane behind recovers and becomes ready to be stimulated again.

Excitable Cells

Signal-carrying animal cells include neurons and muscle cells. In neurons, the charged cell membranes send a signal from one end of the cell to the other. In muscle cells, as a signal is transmitted, this is accompanied by a twitching of the cell, making the muscle contract. In each case, the nature of the electrical charge is much the same.

Neurons and muscle cells perform their extraordinary feats because they are excitable. This means that their charges flip when they are stimulated, such as by touch or exposure to a chemical. The flip occurs because the stimulus affects the charge-generating proteins embedded in the cell membrane. These proteins are extraordinarily sensitive, and a slight disturbance will alter what they do. As a result, their usual routine, making positive charges flow out, is changed. Positive charges instead are blocked and accumulate on the inner surface of the membrane, and the charges are flipped around. A stimulus has made the inner surface positive and, in comparison, the outer surface negative.

The charge reversal doesn't stop there. The sensitive membrane proteins are scattered along the length of the cell, and this causes a domino effect.

The charge reversal stimulates more proteins that are nearby, and very quickly the charge reversal zips right along the neuron from one end to another. In other words, the cell becomes "excited". In nerve fibres that are insulated with a myelin sheath, the charges are concentrated at the gaps, called "nodes", between one insulating Schwann cell and the next. This makes the signal "jump" from node to node to make things even faster. All this would be enough to explain the lightning-fast movements

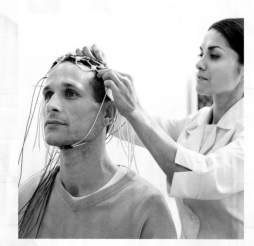

Measuring a Brain's Electrical Activity

Although the charge differences across individual neurons are tiny, they are measurable, especially when the combined effects of many neurons in the nervous system are involved. A technique called electroencephalography (EEG) is used to detect and record this electrical activity in the brain. It involves placing sensors called electrodes on the scalp, while an electronic device, called an electroencephalogram, records the activity. The technique can examine what is happening in the brain when the body is being stimulated, or when the brain generates its own rhythmic patterns of activity: so-called brain waves. EEG is used to help diagnose various problems in electrical activity of the brain, such as epilepsy and sleep disorders.

of any signal, but there is a problem. If the entire membrane suffers a reversal of charge in this way, surely that would block any further stimulus from having an effect? The nervous system has an answer to this problem: the membrane proteins can recover their original state.

Cell Recovery

Very quickly after charge-moving proteins have flipped by being stimulated, they automatically go back to their original state, and the membranes return to their unexcited states. Neuron stimulation typically starts at one end of a nerve fibre, so the movement of excitation can only happen in one direction: away from the tip. As the zone of excitation moves further down the fibre, the proteins behind start to recover, so the membrane returns to its unexcited, or "resting", state. This explains two things. First, a zone of excitation moves down a fibre in one direction only (in other words, it can't excite proteins behind it and move back again). Second, once the zone of excitation reaches the other end of the neuron, the entire neuron can recover, so it is ready to be stimulated again.

The zone of excitation that zips along the membrane of a neuron is called a nerve impulse. It relies on moving electrical charges, which can move very fast: the speed of a travelling nerve impulse is more than sufficient to explain the quick responses of animals. Typical speeds of nerve impulses range from 1–100 metres (3–300 ft) per second. That's equivalent to 3.6–360 kilometres per hour (2.2–220 mph). Given that the biggest animal that has ever lived (a blue whale) grows up to 30 metres (100 ft) long, these figures alone would suggest that it takes just a fraction of a second for an impulse to get from nose to tail in the largest animal of all. But there is one final problem: single neurons cannot span such distances, and nerve impulses can only travel as far as the membranes that carry them. When impulses reach the end of a nerve fibre, there is a tiny gap, called a synapse, between that neuron and the next. And a different signalling technique is needed to cross it.

How Nerve Cells Work

How Nerve Cells Communicate

When a nerve impulse arrives at the end of a nerve fibre, it triggers the release of a chemical to cross the gap between two nerve cells. This stimulates the end of the next nerve fibre. The gaps and the chemicals play a key role in control and coordination.

Your complete nervous system is made up of billions of neurons, many of which are in the brain. Even the simplest animals such as jellyfishes, which don't even have a brain, have a bewilderingly large number of neurons, giving their behaviours a huge amount of potential complexity. Lots of neurons means lots of possible pathways for signals to move. As we have seen, a single neuron, especially one from a brain, is armed with fibres that radiate away from the cell body in different directions. Each fibre can pass its signal to any one of the many fibres coming from an adjacent neuron. This means that, when a nerve impulse moves along a neuron, it has the option of stimulating lots of neuron neighbours.

The actual route the signal takes will determine the outcome of the behaviour. In a human brain, with all its tangled neuronal connections, the possibilities look endless.

But between one nerve fibre and the next, a tiny but significant obstacle lies in the way of any signal: the gap between the end of one fibre and the beginning of the next. This gap is the synapse, and it is no more than 40 nanometres across. But that's ten times wider than the usual gap between adjacent body cells, and enough to prevent a nerve impulse's effect from jumping across. For a signal to pass across the synapse, it needs to change from an electrical signal to a chemical signal.

Neurons communicate with one another by passing chemical messages across the synapse.

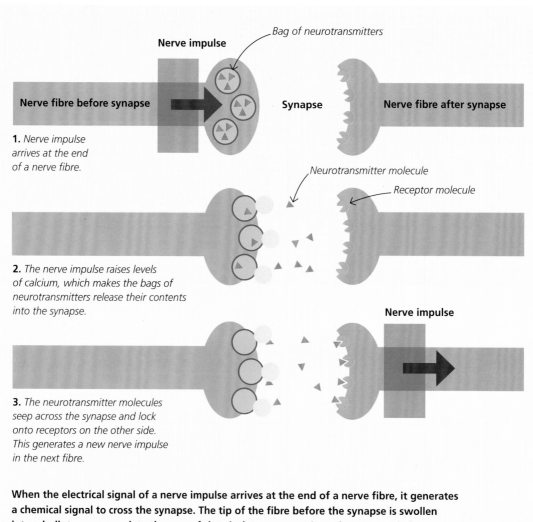

Nerve impulse

Bag of neurotransmitters

Nerve fibre before synapse

Synapse

Nerve fibre after synapse

1. *Nerve impulse arrives at the end of a nerve fibre.*

Neurotransmitter molecule

Receptor molecule

2. *The nerve impulse raises levels of calcium, which makes the bags of neurotransmitters release their contents into the synapse.*

Nerve impulse

3. *The neurotransmitter molecules seep across the synapse and lock onto receptors on the other side. This generates a new nerve impulse in the next fibre.*

When the electrical signal of a nerve impulse arrives at the end of a nerve fibre, it generates a chemical signal to cross the synapse. The tip of the fibre before the synapse is swollen into a bulb to accommodate the sacs of chemical neurotransmitter that are needed.

Chemicals Crossing the Gap

When a nerve impulse arrives at the tip of a nerve fibre, it has nowhere else to go. But before the fibre tip returns to its unexcited state, it needs to pass its signal to the next nerve fibre waiting across the synapse. The membrane proteins that were involved in generating the zipping nerve impulse are now involved in passing on the signal, but different kinds of membrane protein are called into play.

First, special proteins at the tip of the fibre are triggered to fill the tip with calcium. Calcium

is a common trigger in excited cells, and is involved in making muscle cells contract. But at a synapse, it is needed just for sending out the chemical signal. It does this by activating tiny bags of signalling chemicals called neurotransmitters. These bags have been waiting in the tip of the nerve fibre, but upon the arrival of the impulse, and triggered by the sudden rise of calcium, the bags begin to move towards the edge of the cell. Once there, the bags release their neurotransmitters into the synapse.

How Nerve Cells Work

Once liberated, the neurotransmitters seep across the synapse between the two cells towards the membrane of the fibre on the other side. The membrane of this fibre is packed with a kind of protein that intercepts them. These proteins are called receptors. A different kind of receptor is needed for each kind of neurotransmitter: when two complementary partners come together, they fit like the pieces of a jigsaw puzzle. The pairing stimulates the second nerve fibre to generate a new nerve impulse, which continues on its way down the fibre of this new neuron.

In the nerve net of a jellyfish and other simple related animals, signals can pass both ways across a synapse. In more complex animals, the synapses are asymmetrical. With a few exceptions, only one side can ever generate bags of neurotransmitter and only the other has the receptors for grabbing it. This means that signals moving from one neuron to the next can only move in one direction.

How Synapses Recover

As with an impulse-carrying nerve fibre, it is important that a synapse recovers so that it can function again when another impulse arrives. This means that the neurotransmitter must have only a temporary effect. If its effect lasted too long, the receptors in the next neuron would carry on being stimulated by neurotransmitter, and the entire nervous system could get over-excited. Synapses have a way of ensuring this does not happen.

When the first neurotransmitter molecules have locked onto the receptors long enough for an impulse to be generated, the second nerve fibre might respond by sending out a new kind of chemical (an enzyme) to break any excess neurotransmitter down. Alternatively, other membrane proteins may draw the neurotransmitter out of the synapse. Either way, the synapse quickly recovers. And while new neurotransmitter is made in the fibre tip, the system is ready to fire a new signal again.

Neurotransmitters With Different Effects

Many neurotransmitters work in the way described here, generating a nerve impulse in the next nerve fibre, but some have precisely opposite effects. They inhibit the next nerve fibre, making it less likely to respond to being stimulated. The balance between neurotransmitters that excite and others that inhibit is crucial to allow the nervous system to control behaviour. Not only do neurotransmitters work differently, the same one can even have different effects in different parts of the body.

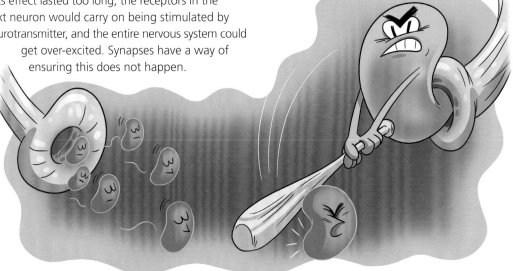

Various mechanisms ensure that a neurotransmitter only has an instantaneous effect, and that it doesn't linger in the synapse to overstimulate the next neuron.

The neurotransmitter serotonin performs many different tasks. It affects the part of the brain that regulates the cycle between sleep and wakefulness. Higher levels of serotonin can also make you feel sick.

An important neurotransmitter called acetylcholine is generally excitatory: it stimulates new nerve impulses in neighbouring nerve cells and makes the muscles that are under conscious control – those connected to the skeleton – twitch. But it inhibits the muscle of the heart, making its beating rate slow down.

Neurotransmitters in the brain are just as important in explaining how the brain works as the electrical signals. And here, their levels and balances influence everything from conscious movement to learning and moods. Dopamine, for instance, generates feelings of pleasure, as well as being important in controlling movements. (Insufficient dopamine results in Parkinson's disease.) Among other things, serotonin regulates moods and sleep, while endorphins reduce levels of stress. Overall, sensations generated by the brain depend on a complex interplay between all these neurotransmitters and more besides.

How Drugs Can Interfere With Neurotransmitters

Certain chemicals can interfere with the way synapses and neurotransmitters work, and many of these effects help to explain the way certain drugs or poisons work on the brain. Some of these chemicals reinforce the effects of neurotransmitters: they are said to be agonistic; others counteract the effects: they are antagonistic. Nicotine, for instance, is an agonist of acetylcholine, the important exciter of the nervous system. This helps to explain its effects as a stimulant. It does this by mimicking acetylcholine and locking onto its synapse receptors, so it ends up doing much the same thing. But antagonistic chemicals can lock onto receptors and block them instead. Curare is a poison found in certain plants, which blocks receptors on nerve cells and muscles, causing paralysis. Antagonists of brain neurotransmitters can be used to treat disorders. Dopamine agonists and antagonists, for instance, are used as antipsychotic drugs to treat a range of conditions, including depression and bipolar disorder.

Chapter 3
HOW THE BRAIN IS CONNECTED

The Spinal Cord

Connecting directly to the base of the brain and running through the spine, the spinal cord is the principal route by which nerve impulses pass up and down the body.

The brain is connected to the spinal cord, and together the two make up the central nervous system. They are both made of neurons that carry bioelectric impulses, and they both process information from the body's sensors, so they can send out impulses to body parts such as muscles that make the responses. But while brain processing includes "higher level" business such as conscious thought, the spinal cord is mostly a conduit for ferrying information from place to place.

The spinal cord may be little more than a conduit, but it serves as the critical link in the nervous system. Practically all the system's nerve signals from the neck down must go through the spinal cord, and for most of our body, its connection to the brain is the only way our thoughts can be turned into actions. The spinal cord's importance was recognized more than three millennia ago. Ancient Egyptian texts describe the devastating consequences of a spinal cord injury:

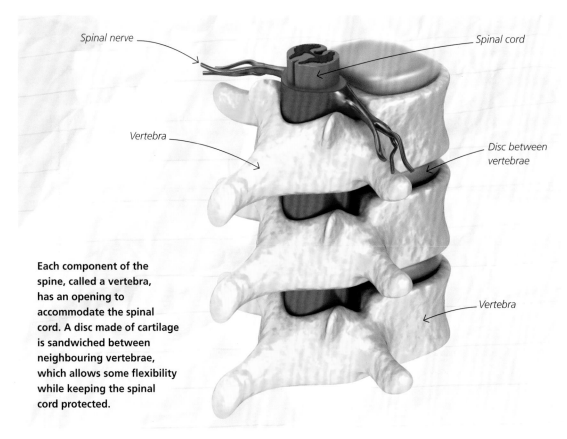

Spinal nerve

Spinal cord

Vertebra

Disc between vertebrae

Vertebra

Each component of the spine, called a vertebra, has an opening to accommodate the spinal cord. A disc made of cartilage is sandwiched between neighbouring vertebrae, which allows some flexibility while keeping the spinal cord protected.

How the Brain is Connected

"a dislocation of a vertebra of his neck extending to his backbone, which cause him to be unconscious of his two arms and his two legs." In the second century CE, the ancient physician Galen, famous for his pioneering work on anatomy, discovered, in a series of gruesome experiments, that if live pigs had their spinal cord cut at different places, it paralysed them at different levels of their body.

On average, the human spinal cord in an adult human body is just less than half a metre (18 in) long, running from the brain to the lower back, and ends at a point a short distance below the level of the rib cage. It is less than a finger's width in diameter and is thickest in the neck. Just as the brain is protected inside the skull, so the spinal cord is protected inside the vertebral column, or backbone. Along its length there are branches: the spinal nerves. These carry electrical impulses to the spinal cord from the body's sensors, and from it to the muscles and glands.

Charles Bell (1774–1842) produced several works on the anatomy of the brain. A skilled artist, he illustrated the books himself.

A Superhighway for Nerve Impulses

The human backbone contains 33 bones, called vertebrae, and the spinal cord runs through an opening in each one. The spinal cord reaches down as far as the top of the lumbar (lower back) region, where it is replaced by a bundle of spinal nerves that continues down to the coccyx, the bony bit at the base of the spine. Above this, there are pairs of spinal nerves that, with a couple of exceptions, roughly correspond to the positions of the vertebral bones. In total, there are 31 pairs of spinal nerves going to left and right sides, including the ones that are bundled in the lower back. Each pair is concerned with ferrying impulses to and from different parts of the body. For instance, the fifth pair from the top controls the movement of the arms, while those a bit further down in the chest region help to control muscles for breathing.

As we saw in chapter 2, the brain and spinal cord of the central nervous system are packed with impulse-carrying cells called association neurons. These have the job of processing the information between receiving impulses from sensory neurons and sending out impulses along the motor neurons.

But the neurons are not packed randomly: they are arranged to make specific circuits so they can not only transfer information from sensors to motor neurons, but also up and down the spinal cord to the brain and back. Many of the neuron fibres inside the spinal cord are organized into bundles called tracts. Ascending tracts carry impulses up the spinal cord to the brain; descending tracts carry them down, away from the brain.

Like all nerves, the spinal nerves contain the fibres of sensory and motor neurons. These are also arranged and connected to the spinal cord in a particular way. The nineteenth century Scottish surgeon Charles Bell worked out how. In a series of vivisection experiments that were reminiscent of those of Galen, Bell found that he could provoke a rabbit's muscles to twitch if he touched the underside of the spinal cord, but not the upper side. He demonstrated that only the underside was connected to the motor neurons in the spinal nerves, leaving the upper side to receive signals from sensory neurons. Today, Bell is perhaps better known for discovering the disorder of nerves connected to the brain that causes paralysis of facial muscles, which is named after him: Bell's palsy.

Sensory and Motor Circuits

Within the protection of the vertebrae, as Bell demonstrated, the body's sensory signals arrive at the spinal cord through the back, or dorsal, side. The motor signals (leading to muscles and glands) leave through the lower, or ventral, side. Each spinal nerve carries a mixture of sensory and motor fibres, and this means that the nerve must split into dorsal and ventral roots where it joins the spinal cord. (This simple two-way split does not happen in nerves of the head that connect directly to the brain: the so-called cranial nerves described in the next section.) The dorsal roots carry fibres of sensory neurons, which connect via synapses with the association neurons in the spinal cord. Other synapses make the connection with the motor neurons that lead out through the ventral root. This circuitry takes the impulses coming from sensors through a pathway that leads to muscles and glands, but does not on its own allow communication with the brain. For this to happen, additional connections between association neurons run through more circuits going up and down through the spinal cord.

Grey and White Matter

As we saw in chapter 2, parts of the brain and spinal cord appear darker or lighter depending upon the distribution of the neuron's cell bodies and fibres. A cell body contains the cell's nucleus and most of its cytoplasm, but nerve fibres are long spindly threads of cytoplasm, some of which may be coated in a fatty layer of insulation, called a myelin sheath. A mass of many cell bodies appears as darker grey matter, while bundles of nerve fibres make up white matter. White matter is therefore concerned with ferrying signals. Grey matter, packed with lots of communicating synapses,

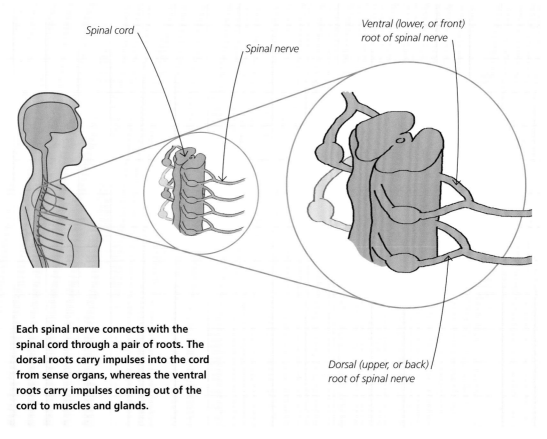

Spinal cord

Spinal nerve

Ventral (lower, or front) root of spinal nerve

Dorsal (upper, or back) root of spinal nerve

Each spinal nerve connects with the spinal cord through a pair of roots. The dorsal roots carry impulses into the cord from sense organs, whereas the ventral roots carry impulses coming out of the cord to muscles and glands.

How the Brain is Connected

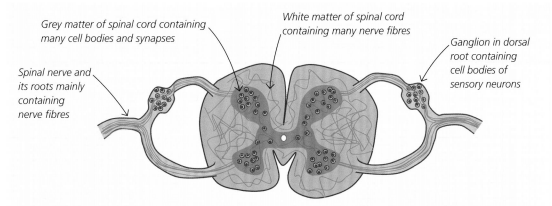

Grey matter of spinal cord containing many cell bodies and synapses

White matter of spinal cord containing many nerve fibres

Ganglion in dorsal root containing cell bodies of sensory neurons

Spinal nerve and its roots mainly containing nerve fibres

A section through the spinal cord reveals a darker inner X-shaped region of grey matter containing most of the cord's cell bodies and synapses. The rest of the spinal cord's white matter mainly consists of nerve fibres, carrying impulses to and from sensors and responders or to and from the brain.

is involved more in processing information. The grey matter of the spinal cord is packed inside its core, where it processes signals coming to and from the sensors and muscles. The grey matter is surrounded by white matter, which mainly contains the long fibres of the tracts that carry signals to and from the brain. Grey and white matter of the brain, as we shall see, is arranged somewhat differently.

There is no distinction between grey and white matter within nerves, because they mainly consist of nerve fibres. But in places, bunches of cell bodies appear as swellings called ganglia (singular ganglion). There are ganglia in the dorsal roots of the spinal nerves because this is where the cell bodies of their sensory neurons are concentrated.

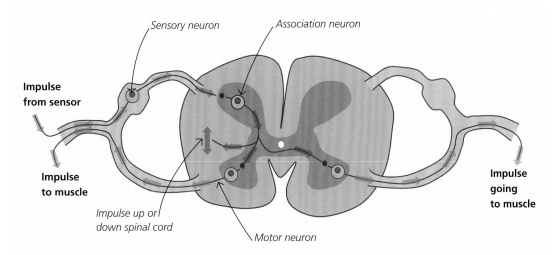

Sensory neuron

Association neuron

Impulse from sensor

Impulse to muscle

Impulse up or down spinal cord

Motor neuron

Impulse going to muscle

Red arrows show the pathway taken by nerve impulses that enter the spinal cord from a sense organ of the body along sensory neurons. Inside the spinal cord, impulses can then pass through different possible routes, shown by purple arrows. Impulses emerge from the spinal cord through motor neuron pathways, shown by green arrows.

The Spinal Cord

How the Spinal Cord Communicates with the Brain

Tracts of neurons running through the spinal cord connect with neurons inside the brain. Some tracts carry impulses into the brain; others carry them away from it. This method of communication provides a link between the "primitive" functions of the spinal cord and the "higher" parts of the brain involved in controlling behaviour. It allows you to be consciously aware of what is going on around the body and gives the brain a say in how sensory stimuli translate into responses. Perhaps most of the information that is processed by the spinal cord in this way is passed up to the brain for evaluation. The brain calculates the sort of response that needs to happen following a stimulus.

Many ascending tracts pass their impulses to a region deep inside the brain called the thalamus. The thalamus is a central processing hub for most of the body's sensory information. Impulses coming from all around the body, including sensors of the skin involved in touch, temperature-detection and pain, end up passing up through there. (As we shall see in the next section, impulses from eyes and ears take a more direct route to the thalamus through the brain's cranial nerves.) The thalamus then distributes impulses to the higher parts of the

The knee-jerk is a simple automatic reflex response that involves impulses passing through the spinal cord. We are consciously aware of the movement, but the system works so fast that the brain doesn't get a chance to interfere.

brain involved in conscious decision-making. Meanwhile, descending tracts carry impulses down from the brain to motor neurons to trigger contraction of muscles in limbs. This entire system brings movement under conscious control.

Other tracts carry impulses upwards from sensors, but do not use conscious parts of the brain. Instead they link to parts that work automatically, without conscious decisions. For instance, coordination of complex body postures and movements, such as

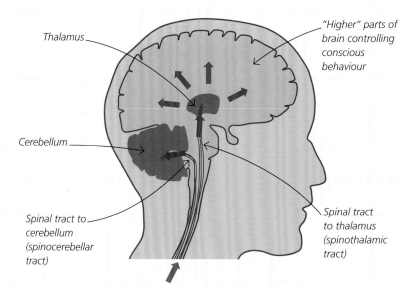

The thalamus and cerebellum coordinate sensory information from around the body. The impulses generated by sensors are carried up to these parts of the brain in the spinal cord (purple arrows). The cerebellum controls automatic unconscious responses, while the thalamus communicates with other parts of the brain that make conscious decisions about responses.

Thalamus

Cerebellum

"Higher" parts of brain controlling conscious behaviour

Spinal tract to cerebellum (spinocerebellar tract)

Spinal tract to thalamus (spinothalamic tract)

The level of a spinal cord injury critically determines the extent of its impact on the rest of the body. This is because injuries at different levels disrupt the flow of impulses through specific sets of spinal nerves to certain muscles. C, T and L injuries occur in different sections of the vertebral column: cervical, thoracic or lumbar.

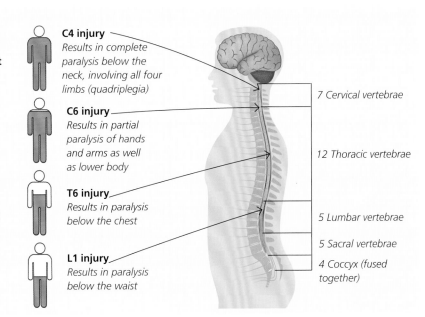

C4 injury
Results in complete paralysis below the neck, involving all four limbs (quadriplegia)

C6 injury
Results in partial paralysis of hands and arms as well as lower body

T6 injury
Results in paralysis below the chest

L1 injury
Results in paralysis below the waist

7 Cervical vertebrae

12 Thoracic vertebrae

5 Lumbar vertebrae

5 Sacral vertebrae

4 Coccyx (fused together)

walking or even just standing upright, is under the control of a part of the brain called the cerebellum (see chapter 5).

When the Spinal Cord Overrides the Brain

Some responses involve no intervention by the brain at all, even though the conscious parts of the brain are aware of them. These are reflex responses that are rapid and automatic. Muscles, for instance, contain bundles of sensors called spindles, which fire off impulses when they are stretched. To prevent possible damage to the muscle, an automatic reflex stops them from being stretched too far: the impulse passes into the spinal cord along sensory neurons, then down through motor neurons back to the muscle to make the muscle contract. A well-known example of an especially simple kind of automatic response is the knee-jerk reflex, where tapping the knee of a seated subject triggers the leg to jerk upwards. Stretch reflexes like this can bypass the brain entirely because the sensory neuron synapses connect directly with the motor neuron. The pathways involve no association neurons, meaning that a signal cannot be sent up to the brain for interception.

How Injury Reveals Function

Sometimes the biggest clues about the workings of the body show up when bits of the body go wrong. Scientists can learn by studying patients with injuries or diseases that affect specific parts of the body. Such studies have been particularly revealing when it comes to the brain and nervous system, especially in the days before techniques such as electro-stimulation were perfected to test the body more experimentally. For instance, damage to the grey matter in the dorsal part of the spinal cord causes loss of sensation because this is where sensory signals are received, but leaves motor control of muscles intact. Damage to the ventral part will disrupt motor function but not sensation. These kinds of injuries corroborated the previous findings of Charles Bell. What is more, the parts of the body that are affected depend on the section of spinal cord involved. In the brain, local damage similarly points to local function. For instance, blindness results from damage to the occipital lobe at the back of the brain, showing that this region is involved in vision.

Wiring of the Brain

Inside the head, the brain connects directly with sense organs through cranial nerves. Most of these link to the facial senses of the eyes, ears, mouth and nose, while additional wiring inside the brain coordinates its different parts.

Although scholars in ancient Egypt and Greece knew that damage to the brain or spinal cord affects what we can do or how we think, not everyone was convinced the brain itself was the body's ultimate control centre. Among the Greek philosophers, Hippocrates and Plato thought that the brain was the seat of mental activity (Hippocrates even linked epilepsy to a disturbance of the brain), but Aristotle thought that this was something that happened in the heart. According to Aristotle, the brain, with its rich blood supply, was used to dispel excess heat, like a radiator. Aristotle's views were debunked, of course, but they persist in our everyday sayings: we can still "think with our hearts" and "keep a cool head".

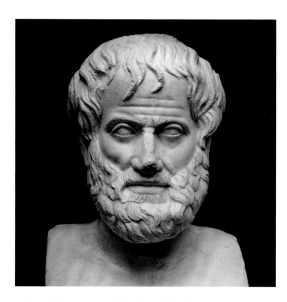

Aristotle (384–322 BCE) believed that humans' large brains helped us to cool off and keep our tempers under control better than smaller-brained animals.

The very fact that the brain is connected to the spinal cord was obvious to anyone who dissected a body. But early anatomists would also have seen how the brain connects directly to the sense organs of the head: nerves link the brain to the eyes, ears, nose and tongue. In fact, just like the spinal cord, the brain has motor connections, such as those to the muscles that control our facial expressions. The nerves attached to the brain are called cranial nerves. Humans have about 12 pairs of cranial nerves (the exact number depends on how they are defined) and, just like the 31 pairs of spinal nerves, they have a very precise arrangement. Some cranial nerves only contain sensory nerve fibres or motor nerve fibres, while others have a mixture of both.

Cranial nerves allow the brain to communicate directly with the sensors and muscles of the face and head. But to work properly, the brain also needs an internal communication system that works from within. This system exists as neuron circuits that help one part of the brain interact with another. The interconnections depend on precisely arranged regions of grey and white matter.

Cranial Nerves

The plan of the cranial nerves reads like a plan of the face and head. From the front, the first pair of cranial nerves are sensory nerves that send impulses from the nose for sensing smell. The second, third, fourth and sixth pairs connect with the eyes and include, first, the sensory nerves for sight (the optic nerves) and then motor nerves for moving the eyeballs. The fifth pair of cranial nerves contain a mixture of sensory and motor fibres that link to sensors and muscles of the face, while the seventh is also a mix, of sensory fibres from taste buds and facial muscles. The eighth cranial nerves have sensory fibres from the inner ear for sensing sound and balance. The ninth, a mix of sensory and motor, connect to taste sensors and muscles at the back of the throat. The eleventh and twelfth cranial nerves contain just motor fibres that connect to muscles of the neck and tongue respectively. That leaves the tenth cranial pair. This is called the vagus nerve, whose name comes from the Latin vagus, meaning "wandering". It travels widely and branches to serve parts of the mouth, throat and vital organs of the chest.

In the chest, the vagus nerve is most important for helping to control the heart, lungs and other vital organs. It connects to a "primitive" part of the brain that is concerned with their regulation, called the medulla oblongata (see chapter 6). It also sends nerve impulses to muscles of the heart to slow down heart beat rate. (Another nerve coming from the spinal cord has the opposite effect of speeding it up.)

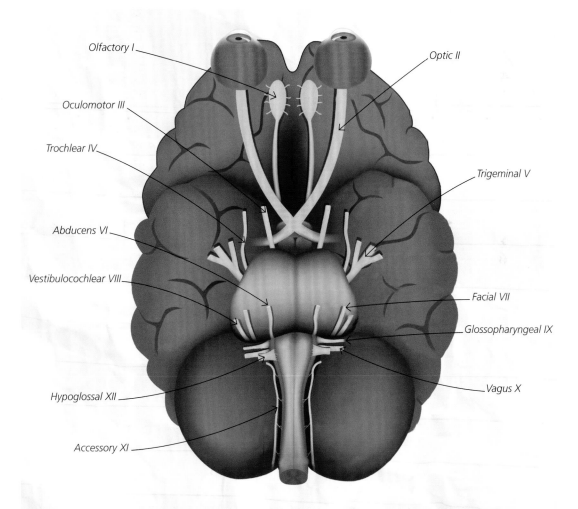

Olfactory I
Optic II
Oculomotor III
Trochlear IV
Trigeminal V
Abducens VI
Vestibulocochlear VIII
Facial VII
Glossopharyngeal IX
Vagus X
Hypoglossal XII
Accessory XI

Twelve pairs of cranial nerves connect to the underside of the brain. Some carry impulses from sense organs, some carry impulses to muscles and others carry a mixture of both.

Together with the layers of grey matter on the brain's surface, there are very many centres of grey matter, called nuclei, deeper inside the brain that are responsible for controlling different tasks.

Centres of Brain Activity

Although the outer shape of the brain, including its bulges and folds, gives a hint about the way it manages what it does, there is a deeper, hidden organization inside the brain. This organization depends on the arrangement of neurons and neuron fibres lying far beneath the bulges and folds.

The neurons of the brain are grouped into clusters called nuclei (not to be confused with the microscopic nuclei that contain DNA inside individual cells). A brain nucleus has a particular size and shape, and is usually distinguished as a patch of grey matter surrounded by white. Some nuclei are smaller than a grain of rice; others are much bigger. And some are coloured. For instance, the red nucleus is so-called because its high iron content gives it a tinge of pink. It controls aspects

of motor coordination, such as helping babies to crawl. Other nuclei are similarly specialized in what they do, often to a very fine degree. The regulation of heart rate is not only restricted to a part called the medulla oblongata, but more precisely localized to specific clusters of brain cells that have antagonistic effects: one cluster speeds the heart up, while another slows it down. Other parts of the brain have their own nuclei, meaning that the brain is precisely organized with centres within centres.

The thalamus, as we have seen, is involved in processing incoming sensory information. It has nuclei specifically concerned with receiving visual signals, storing memories of sensations, etc. Altogether, there are many nuclei in the human brain, each one specialized to perform a very

How the Brain is Connected

particular task within the entire complex business of the working brain. Many are found in the brainstem; others are deep within each cerebral hemisphere – the two half-domes that make up most of the "higher" parts of the brain.

Not all complex brain activities rely on these kinds of nuclei. On the surface of the cerebral hemispheres there is a particularly rich gathering of grey matter that extends over the wider area. It is responsible for some of the most complex "higher-level" brain activities. This so-called cerebral cortex is separated from deeper nuclei of the brain by tracts of white matter that carry impulses between the two. It contains lots of cell bodies and synapses, but its grey matter is not organized into nuclei. Instead, the cerebral cortex, which is critical to perception, awareness, thought and consciousness, consists of layers of grey matter. So much is needed for the

sophisticated business of conscious thought that it must be folded beneath the skull, creating the convolutions on the surface of the human brain.

The Unconscious and Conscious Brain

With its deep nuclei and folded cerebral cortex, the brain continually performs tasks that we take for granted. These tasks range from the routine, such as changing heartbeat or breathing, to the idiosyncratic, such as determining personality and mood swings. And although the deep nuclei of the brain are organized in an exceedingly complex fashion, the outer shape of the brain does, at least, give a big clue as to how all these tasks are organized. For a start, there is a distinct difference between the brainstem and the paired cerebral hemispheres: the half-domes.

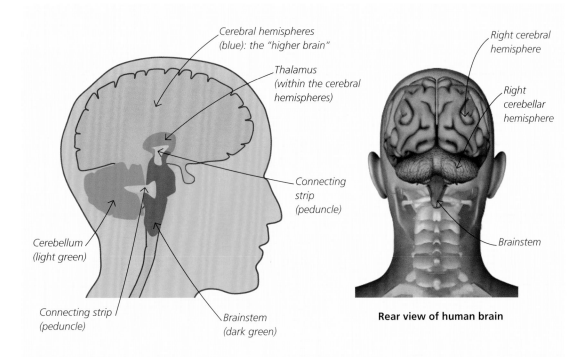

Cerebral hemispheres (blue): the "higher brain"

Thalamus (within the cerebral hemispheres)

Connecting strip (peduncle)

Cerebellum (light green)

Connecting strip (peduncle)

Brainstem (dark green)

Right cerebral hemisphere

Right cerebellar hemisphere

Brainstem

Rear view of human brain

The simplest plan of the brain distinguishes the automatic control centres of the brainstem and cerebellum from the higher conscious regions in the cerebral hemispheres. The brainstem forms a single central stem of the brain; all other parts shown here are paired.

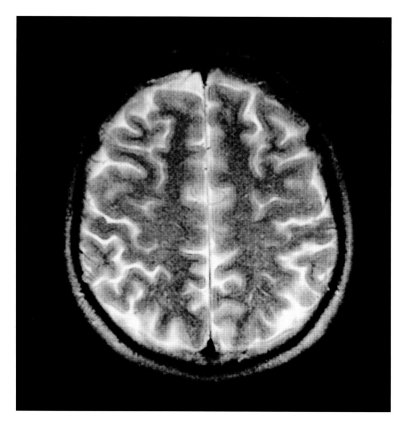

Seen from above in this MRI scan, the cerebral cortex is a layer about 2.5 mm ($^1/_{10}$ in) thick on average, covering each hemisphere of the cerebrum. Here, neurons constantly form new connections with one another as we think, decide and act.

The brainstem is an extension of the spinal cord, reaching up into the skull, where it connects with the base of the cerebral hemispheres. It controls some of the most basic functions needed to keep us alive. Activities such as the heartbeat, breathing, and the flow of hormones are controlled by parts of the brainstem. We have no conscious control over these activities, which take place automatically without us having to think about them. In simple terms, the brainstem is the unconscious part of the brain. It widens towards the top, where it splits into two broad strips (called peduncles): one strip goes to one cerebral hemisphere, one goes to the other. These strips contain the major routes for impulses to pass to and from the hemispheres. Right at the top of each strip is one of two lobes of the brain's thalamus: the principal hub for processing sensory information (see previous section).

The cerebral hemispheres are not the only paired structures that are obvious from the overall shape of the brain. About midway along the brainstem there are more strips extending towards the rear of the brain, which connect to the paired parts of another folded region of the brain: the cerebellum. As we saw in the previous section, the cerebellum coordinates complex movements. These movements can be brought under conscious control, but, like the tasks controlled by the brainstem, the cerebellum does a lot of its work without conscious control. However, in their structure, the hemispheres of the cerebellum more closely resemble the cerebral hemispheres: they have a folded surface of grey matter.

The cerebral hemispheres themselves, in the human brain at least, are bigger than all the other brain parts put together. This reflects their importance in controlling complex tasks: decision making, storing memories, controlling mood and even personality. They make up the "higher" parts of the brain, both in terms of position and in what they do. These parts of the brain give us a sense of control over our actions.

Chapter 4
INSIDE THE BRAIN

Brain Cells

An average human brain contains more than 100 billion cells. Many of these are supporting cells that nourish and protect but do not carry bioelectric impulses. Only the neurons are responsible for the brain's thought processes.

The adult human brain is about 17 cm (7 in) long, 14 cm (5½ in) wide and 9 cm (3½ in) high. It weighs 1.2–1.4 kg (2 lb 10 oz–3 lb 2 oz) and makes up about 2 per cent of the weight of the entire body. The brain of a newborn baby weighs about 400 grams (1 lb), so about 1 kg (2 lb) of weight is added during growth into adulthood. A fresh brain is a mixture of white, grey, black and red, but its colours fade and become yellowish after storage in preservative. It is soft and dense but squishy, with a texture that has been variously described as like jelly or a ripe avocado. Making up this consistency are the billions of microscopic cells

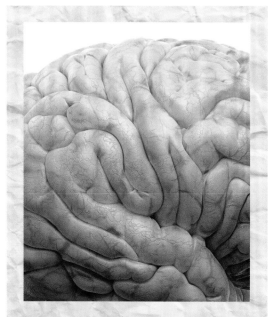

The brain has a squishy consistency. If you push into it, you will create a dent, which springs back when you let go.

that perform every job that gives you intelligence, emotion, a personality and keeps you alive.

Many of these cells, of course, are neurons that carry nerve impulses and communicate with each other through chemical synapses, just like all the other neurons of the nervous system. But the brain also contains other kinds of cells that are necessary to help keep the brain alive: to protect, support and nourish.

Life-supporting Cells

The cells of the brain that are not neurons are called "glial" cells, from the Greek word meaning "glue". But these cells do much more than just keeping brain parts stuck together. Although some do seem to hold the neurons in place, most glial cells do more complex tasks. Some help to supply neurons with food and oxygen, or keep the brain free of infection-causing microbes and clear up cells that have died. Some glial cells insulate the neuron's fibres with the fatty coatings of myelin sheath that we saw in chapter 2. Glial cells are found throughout the central nervous system, so also occur in the spinal cord.

Like all tissues in the body, brain cells obtain their food and oxygen from the bloodstream. Nutrients such as sugar, fats, amino acids, vitamins and minerals seep into the blood from the digestive system, while oxygen enters the blood from air breathed into the lungs. Food is needed to provide the building blocks to help tissues grow, and some energy-rich materials, such as sugars and fats, are also used as fuel. These materials are used with oxygen during respiration: chemical reactions that release energy. Neurons need a great deal of energy to generate the charge differences on their membranes that they need to send impulses (see chapter 2). Some star-shaped glial cells, appropriately called astrocytes, provide physical links between the hungry brain cells and help to pass them nutrients and oxygen from microscopic blood vessels called capillaries. Astrocytes form an important part of the so-called blood-brain barrier because they help to control what materials can pass from one to the other.

Smaller glial cells called microglia are the brain's front line of defence. They wander between brain cells, carefully checking for signs of infection or damage. Like some white cells in the blood, microglial cells can change shape to engulf foreign particles, such as invading microbes that might cause disease. They also help to control aspects of the body's immune response, for instance by triggering inflammation in response to injury.

Finally, specialized glial cells called ependymal cells line cavities inside the brain and spinal cord. These help to produce the cerebrospinal fluid in the cavities to cushion the central nervous system and ensure that its cells are bathed in surroundings that are regularly cleared of waste. The surfaces of ependymal cells are covered in hair-like structures called cilia, which beat to keep the fluid flowing.

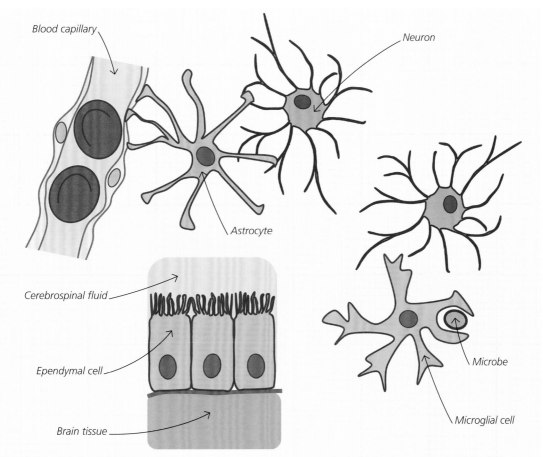

A variety of glial cells help to support the impulse-carrying neurons in the brain. Astrocytes pass them food and oxygen from the blood. Ependymal cells help make and move the cerebrospinal fluid. Microglia destroy foreign particles.

The Brain's Blood Supply

The continuous activity in the brain makes it a hungry organ, and it needs a constant supply of food and oxygen from the bloodstream. Its demand is clear enough from the devastating consequences of any interruption of blood flow. Even a momentary loss of fuel or oxygen, such as from a blockage caused by a blood clot or through internal bleeding, can kill neurons, resulting in a stroke. At the very least, this could leave some kind of physical weakness; at worst, it could result in paralysis or even be fatal.

Blood that has been oxygenated in the lungs is pumped around the body by the heart and reaches the brain through arteries. One pair of arteries, called the vertebrobasilar arteries, supplies the brainstem and lower part of the cerebral hemispheres. The rest of the cerebral hemispheres are supplied by another pair of arteries, called the internal carotid arteries. These arteries run deep inside the neck and connect to a loop under the brain called the circle of Willis. This interconnected arrangement helps to protect the brain's blood supply: if any one artery is interrupted, for whatever reason, blood flow can still reach the brain through the others. Smaller arteries branch from the circle of Willis to supply blood to specific parts

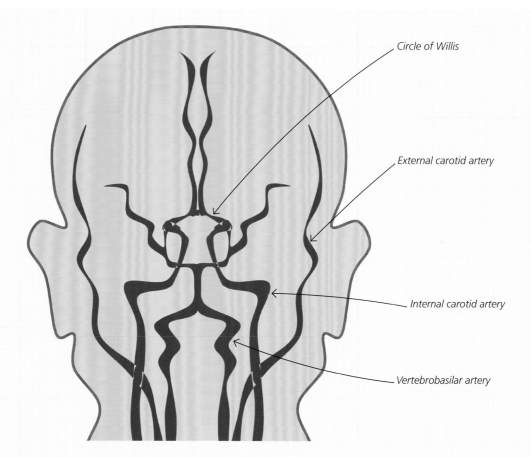

Circle of Willis

External carotid artery

Internal carotid artery

Vertebrobasilar artery

Blood flow to the brain is supplied by four arteries: a pair of vertebrobasilar arteries supplying mainly the brainstem, and a pair of internal carotid arteries supplying most of the cerebral hemispheres. The circle of Willis helps to keep blood flowing if any one artery is obstructed.

Inside the Brain

Can Brain Cells Regenerate?

It was once thought that brain cells could not regenerate: that by age two or three, we already had as many brain cells as we would ever have. From then on, it was thought that brain cells that are lost through ageing, disease or injury would never be replaced. However, recent studies suggest that this might not be the case. In the 1990s, research on monkeys at Princeton University found that new neurons can grow in several parts of the brain. Our bodies contain cells called stem cells, These are the body's undeveloped cells that can become any kind of specialized cell: for instance, we have stem cells in our bone marrow that continually make blood. The Princeton research uncovered neurogenesis (making neurons) from stem cells in the hippocampus – a part of the brain involved in forming memories (see chapter 8) – and also in the cerebral cortex. Later studies on human brains confirmed that the hippocampus can regenerate cells, as well as showing neurogenesis in parts of the brainstem. It is hoped that this research could lead to new treatments for forms of dementia, such as Alzheimer's disease.

of the brain, such as the various lobes of the cerebral hemispheres.

From the arteries, the blood in the brain passes through millions of microscopic blood capillaries, where glucose and oxygen seep into brain cells and waste products pass back. Some of the deoxygenated blood then drains into channels on the surface of the brain, while the rest flows into veins deeper inside. From there, the deoxygenated blood converges through the jugular veins running down the neck, before joining the rest of the body's circulation back to the heart.

How Many Brain Cells?

Although it is generally agreed that the brain contains about 80–90 billion neurons in total, the proportion of glial cells to neurons is not well understood. Recent studies have challenged previous estimates that glial cells outnumbered neurons ten-to-one, but it is still thought that as many as 80 per cent of the brain's cells are supporting glial cells. Most of the neurons of a human brain have developed in the foetus in the first trimester (three months) of pregnancy, but glial cells continue to be added until a few years after birth. And the numbers seem to vary in different parts of the brain too. Grey matter, with its abundance of neuron cell bodies, has proportionately fewer glial cells than white matter. Remember that white matter is little more than

nerve fibres, so it needs plenty of the glial cells that make the fatty myelin sheath. Some regions of the brain, such as the cerebellum (concerned with coordinating movements), seem to have fewer glial cells than the cerebrum (concerned with "higher" functions), but the reasons for these differences are not well understood.

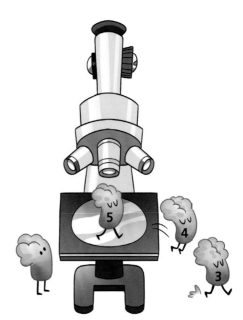

Scientists normally count brain cells by studying slices of brain under a microscope. New, improved methods suggest that the human brain contains about 86 billion neurons, 16 billion of which are in the cortex.

Layers of the Brain

The bulk of the brain is made up of grey and white matter. It is surrounded by protective membranes called meninges, while deep inside it contains fluid-filled chambers called ventricles.

The grey and white matter of the body's central nervous system are both made up of neurons and their supporting glial cells, but they are involved in different aspects of activity. White matter, with its long spindly nerve fibres, some glistening with insulating coats of fatty myelin, is largely used for sending nerve impulses from place to place. Grey matter contains most of the neurons' cell bodies and synapses, and so is the principal site for the complex processes involved in coordination and control. As we saw in chapter 3, in the spinal cord, grey matter forms a central core, which is surrounded by white matter. But within the brain itself, grey matter is also found on the outer lining of the brain, around the cerebrum and cerebellum. This means that many of the complex control functions have shifted outwards from the interior to the surface. White matter has the job of carrying electrical signals between the surface and the interior.

On the very outer surface of the brain there are protective membranes called meninges, while deeper inside there are fluid-filled chambers called ventricles (not to be confused with the chambers of the heart, which have the same name). These outer membranes and inner

The entire central nervous system, including the brain and spinal cord, is surrounded by protective membranes, while grey and white matter contain neurons and supporting glial cells. Deeper in the brain there are chambers called ventricles.

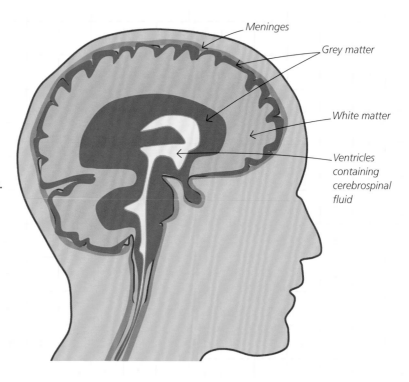

Meninges

Grey matter

White matter

Ventricles containing cerebrospinal fluid

Inside the Brain

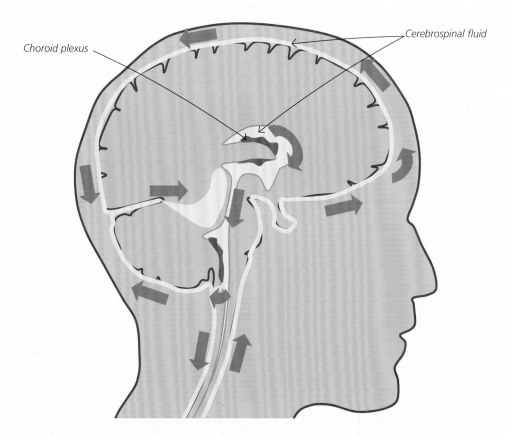

Choroid plexus

Cerebrospinal fluid

Cerebrospinal fluid is produced by patches of blood vessels in the brain called choroid plexuses. From there, the fluid flows through various channels to circulate around the central nervous system through the meninges, wafted along by beating hairs of ependymal cells that line the cavities.

chambers are both bathed in cerebrospinal fluid, which circulates between them.

Cerebrospinal Fluid

The brain is surrounded by a thin layer of cerebrospinal fluid produced from blood. Special patches of blood vessels bordering the ventricles inside the brain filter the blood to make a fluid that ends up being very similar to blood plasma (the liquid part of blood devoid of blood cells), except that it lacks blood proteins. Each patch, called a choroid plexus, passes fluid through a lining of ependymal cells, and the resulting cerebrospinal fluid drains out to fill the brain's ventricles. From

there, it emerges through channels onto the surface of the brain. It also runs down through the central tube and outer linings of the spinal cord. In this way, the entire central nervous system is bathed in cerebrospinal fluid, which cushions the delicate tissues and helps to absorb knocks. The fluid also regulates conditions in and around the brain and spinal cord, flushes out waste products and may even carry chemical signals. It is produced continuously, at the rate of about 500 ml (1 pint) per day, and is continuously reabsorbed back into the blood via veins. At any one time, there is a constant amount in circulation: somewhat less than 200 ml (1 cup) for an average adult human brain.

Brain Membranes

Three membranes surround the brain: these are the meninges (singular meninx). As well as providing a protective coat to the central nervous system, they also contain cerebrospinal fluid and blood vessels. The outer membrane is the most durable, made of tough, fibrous, slightly elastic material that is attached to the inner surface of skull bone. A somewhat loosely fitting middle membrane just below the outer membrane sits on a network of fine filaments that resembles a spider's web. Cerebrospinal fluid flows through the spaces between these filaments. Finally, the innermost membrane is stuck to the surface of the brain itself. It contains the blood vessels that supply nutrients and oxygen to the underlying brain cells.

The spinal cord is also protected by the same set of meninges, but here the durable outer membrane is not fixed to the inner bone surface of the spine. Instead there is a space, called the epidural space, that contains blood vessels and fatty tissue. The epidural space can be targeted for administering drugs, such as local anaesthetics, by a local injection. Going somewhat deeper, it is also

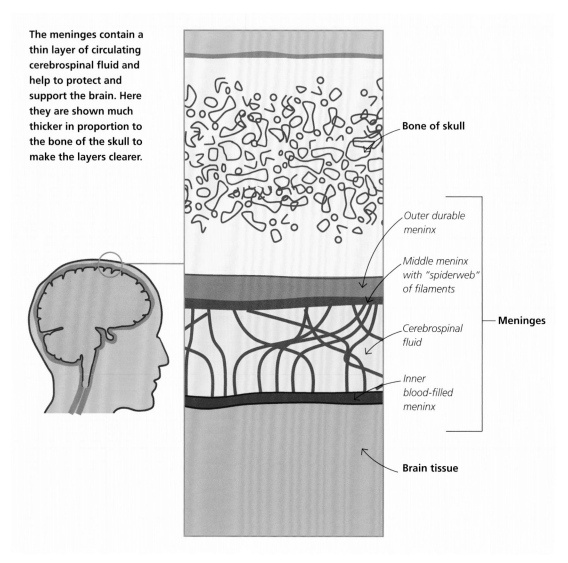

The meninges contain a thin layer of circulating cerebrospinal fluid and help to protect and support the brain. Here they are shown much thicker in proportion to the bone of the skull to make the layers clearer.

Bone of skull

Outer durable meninx

Middle meninx with "spiderweb" of filaments

Cerebrospinal fluid

Inner blood-filled meninx

Meninges

Brain tissue

At the blood–brain barrier, cells called astrocytes prevent harmful substances in the blood from passing into the brain.

possible to take samples of cerebrospinal fluid from the lower back, a procedure called a spinal tap, or lumbar puncture, which involves penetrating the outer meningeal membrane to reach the fluid inside. This is done to help diagnose a variety of medical conditions. Unlike blood, cerebrospinal fluid lacks cells, but injury or disease affecting the brain or spinal cord may cause red or white cells to appear in it.

The Blood–Brain Barrier

Despite the fact that vital substances, such as nutrients and oxygen, must flow between the blood and brain cells, the arrangement of tissues in the brain helps to ensure that potentially dangerous particles, including disease-causing microbes, cannot pass across. The blood–brain barrier is mainly achieved by the specialized glial cells called astrocytes (see previous section). In most parts of the body, useful substances carried inside microscopic blood capillaries pass out through the thin capillary walls directly into the surrounding tissues. But in the brain, the astrocytes act as a "go-between" from blood to vulnerable neurons. The astrocytes control the passage of substances, allowing some through but preventing others, and their principal function is to exclude potentially harmful particles.

When the body is under attack from a disease, inflammation may cause the blood–brain barrier to become more leaky than usual. As elsewhere in the body, inflammation happens to allow infection-fighting white cells through to kill any microbes, such as bacteria or viruses. During this time, the barrier is more likely to be breached by new invaders, but it also may allow drugs, such as antibiotics, to penetrate when they would not ordinarily do so.

Some tissues in the brain, such as the choroid plexuses, form a barrier between the blood and the cerebrospinal fluid but serve a similar function in reducing the risk of infection. However, in other areas of the brain, there is no significant barrier between neurons and blood at all. For instance, parts that are modified as glands (such as the pituitary and pineal glands: see chapter 6) are involved in secreting chemical triggers called hormones into the bloodstream unimpeded.

Meningitis: Breaching the Blood–Brain barrier

Several kinds of microbes, including viruses and bacteria, can pass through the blood-brain barrier and cause infections in the meninges that surround the brain. The result is an infection called meningitis, which causes fever and headache. Bacteria are especially dangerous and may lead to a form of blood poisoning that appears as a skin rash: this happens because they reproduce in the blood, releasing toxins that damage the blood vessels. Bacteria can invade the meninges through the bloodstream, often via the nasal linings.

Research has shown that they can do this because they carry a special kind of protein that allows them to breach the blood-brain barrier. Once in the membranes of the brain, they set up an infection in the layer of cerebrospinal fluid beneath the middle membrane (meninx). The body's natural immune response then causes this region to become inflamed. Meningitis can be fatal if left untreated. It can be cleared up using antibiotics, and long-term protection is now possible by vaccination.

A number of different bacteria can cause meningitis, depending on the person's age. In children, *Streptococcus pneumoniae*, pictured here, is often the culprit.

Chapter 5
A BRAIN OF MANY PARTS

A Plan of the Brain

At a very early stage in the development of an embryo, the main parts of the brain emerge from a swelling at the front of the spinal cord. The parts concerned with controlling vital functions are at the base, while those concerned with complex calculations are at the front.

Before the 16th century, anatomists would dissect brains to study their structure, but progress in understanding how brains worked was slow. They were hampered by deeply embedded traditional beliefs and inadequate technology. Neurons could not be seen without the aid of a microscope, which would be invented in 1590, and science would have to wait until the late eighteenth century to start to appreciate the role of bioelectric impulses, when Italian scientist Luigi Galvani showed that frog legs could twitch when shocked with electricity. In 1504, Leonardo da Vinci dissected a human brain and studied its anatomy by injecting the ventricles (chambers) with wax. He was one of the first to cast doubt on the traditional view that the ventricles were the seats of three facets of mental activity: imagination, reason and memory. Like others before him, Leonardo also searched for, and failed to find, the place of the soul.

It wasn't until the seventeenth century that a modern understanding of the brain's anatomy began to take shape, building on an improved knowledge of physiology (how the body works). Terms likes cerebrum (the umbrella term for the cerebral hemispheres), cerebellum and medulla (the lower part of the brainstem) were used with more consistent meanings. In 1664, English physician Thomas Willis published his work *Anatomy of the Brain*. Willis dismissed the idea of ventricles as chambers for the spirits, correctly deduced that cerebrospinal fluid was produced from the blood, and, by injecting the blood vessels with ink, successfully demonstrated the circular blood flow that still bears his name (the circle of Willis: see chapter 4).

In the following centuries, biologists gained a better understanding of how the different parts of the brain were organized by comparing the brains of animals and studying the way the brain develops in embryos. They found that the tiniest functional human brain, in an embryo smaller than a newly hatched tadpole, had already taken the shape of an adult brain, setting in place the pattern of development for the complex arrangement of bulges, folds and tracts that lie deep inside.

Thomas Willis (1621–1675) identified a host of new structures inside the brain. Detailed drawings of Willis's discoveries were made by Christopher Wren, the architect who designed St Paul's Cathedral in London.

An Embryo Brain in Three Sections

The central nervous system of a human embryo starts off as a cord of neurons stretching down the back, with a thin fluid-filled cavity running through the middle. As this develops into the spinal cord, its front end enlarges to become the brain, and the cavity swells into the ventricles (chambers) of the brain. Now that there is a rudimentary brain, it becomes partially constricted along its length, which divides it into three main sections. From the back to front, these are the hindbrain, midbrain and forebrain. Together, the hindbrain and midbrain will form the brainstem: the "root" of the brain involved in controlling the vital functions. When the embryo's heart starts to beat, it is the brainstem that regulates the pace. Meanwhile, various outgrowths of the brain sprout as the embryo grows bigger. The cerebellum, which coordinates movements, grows backwards from the hindbrain, while the cerebrum, necessary for conscious thought, expands outwards from the forebrain. Long before the baby is born, the basic plan for the brain is set in place. And the hindbrain, midbrain and forebrain are already developing the deeper brain structures.

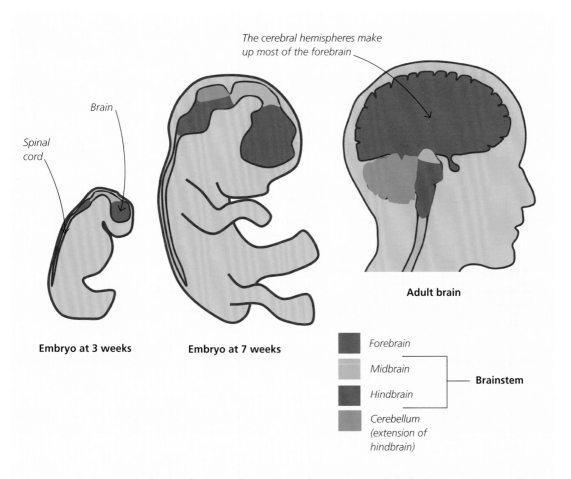

The cerebral hemispheres make up most of the forebrain

Brain

Spinal cord

Embryo at 3 weeks

Embryo at 7 weeks

Adult brain

Forebrain

Midbrain

Hindbrain

Brainstem

Cerebellum (extension of hindbrain)

At a very early stage in the development of an embryo, the main parts of the brain emerge in a swelling at the front of the spinal cord. The parts involved in controlling vital functions are at its base, while those for more complex behaviour are further forward.

Enlargement

Unsurprisingly, animals with the most complex behaviour also have the biggest brains. Behaviour is also reflected in the relative sizes of the parts. But across all backboned animals, the functions of the brain parts are remarkably consistent. Vital functions are typically as challenging in one species as they are in any other: regulation of the heart beat and breathing, for instance, involve similar capabilities in a lizard and a human, and basal parts of the brainstem vary little from animal to animal. But other parts may be highly variable, and this helps to underscore the roles that these parts play in overall brain function. For instance, the cerebellum, the part of the hindbrain that subconsciously controls complex movements, is proportionally large in animals that must coordinate lots of muscles to get from place to place. In birds, it is enlarged for flight. For most fishes, it is the front end of the brainstem (corresponding to a part of the midbrain) that is enlarged. Specifically,

they have an enlarged region called the tectum. This is a part of the brain in backboned animals that helps to process visual information. In fishes, it may also be used as a learning centre. Even though the brain parts of different animals have developed and enlarged in different ways, their central functions usually stay the same. The tectum of fishes that live in dark caves is much smaller, supporting the idea that this area is used by fishes to process visual information. Our tectum is proportionately smaller than that of most fishes, but we also use it to help process vision.

The biggest enlargement in human brains has happened in the cerebrum of the forebrain, divided into the cerebral hemispheres. This is the part involved in the "highest" brain functions, such as problem solving and conscious thought. The cerebrum in primates is larger than that of most other backboned animals, and it is largest of all in our own species, providing us with an abundance of advanced higher-order mental processing power.

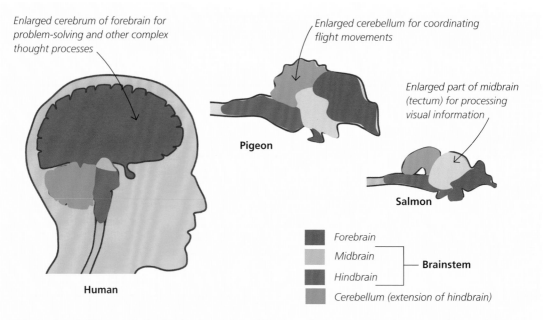

Enlarged cerebrum of forebrain for problem-solving and other complex thought processes

Enlarged cerebellum for coordinating flight movements

Enlarged part of midbrain (tectum) for processing visual information

Pigeon

Salmon

Human

■ Forebrain
■ Midbrain ⎤
■ Hindbrain ⎦ **Brainstem**
■ Cerebellum (extension of hindbrain)

The brains of different kinds of animals are shaped by the enlargement of various parts that perform specific tasks. The brain of each animal has the same parts, but they are developed in different ways according to the animal's evolved behaviour.

The brains of warm-blooded animals, such as birds and mammals, are generally larger and more complex than those of cold-blooded animals, such as amphibians and reptiles.

The Warm-Blooded Brain

When we think of intelligent animals, our minds turn to things like apes and monkeys, whales and dolphins, parrots and crows. These are warm-blooded animals, and it is perhaps no coincidence that they seem to have better brain power than cold-blooded insects, reptiles, fishes and frogs.

Warm-blooded animals have a built-in system for temperature control. Most animals can control their body temperature to a certain extent, but mammals and birds can do it so well that, for most of them, the temperature of their blood stays within a degree or so of an optimum value. In humans, it hovers around 37°C (98°F). That is usually a lot warmer than the surroundings, and helps to keep the cells of the body working to the peak of their performance. The brain is responsible for continually monitoring the blood temperature and commanding the necessary tweaks to bring it closer to the optimum. And it happens independently of the temperature of the surroundings because the body can generate a lot of body heat to warm up, or trigger mechanisms such as sweating or panting to cool down.

The brain of a cold-blooded animal can keep a check on body temperature, but there is no automatic self-correction. On a cold day, a snake must rely on the warmth of the sun to get going. Cold-blooded animals, therefore, slow down in every sense (movement, metabolism and mental processes) when their surroundings are cold, whereas warm-blooded animals can stay active. A brain that is bathed in warm blood all the time has neurons that are continually working at their best. Perhaps this explains why mammals and birds have the biggest brains, and why humans, chimpanzees and parrots appear to have greater brainpower than guppies and geckos.

The Brainstem and Cerebellum

The brainstem ferries nerve impulses between the spinal cord and higher parts of the brain and also controls the workings of vital organs. The cerebellum helps to coordinate complex movements.

The brainstem is often described as the "primitive" brain because it is involved in controlling the most basic vital functions that would have been set in place during the evolution of the first complex animals. Although the heart and lungs (or gills, in water-breathing animals) can work without

French physiologist Marie-Jean Pierre Flourens (1794–1867) studied brain function in experiments with pigeons and rabbits. He destroyed different parts of the animals' brains and observed the changes in their behaviour that resulted.

conscious thought, their activity must still be regulated to ensure that they work at the right pace. The lowest part of the brainstem is the hindbrain. From here, it rises until it merges into the midbrain at the top. In humans, the midbrain is close to the very centre of the brain.

Proof that the brainstem controls vital functions came from experiments on live rabbits conducted in the early 1800s by French physiologist Julien-Jean-César Legallois. Building on experiments showing that a rabbit's spinal cord controlled sensation and movement (see chapter 3), Legallois found that, if the brainstem was cut at the level of the vagus nerve, which connects the brain to vital organs, it stopped the rabbits from breathing. Clearly, without this part of the brain intact, life itself was not possible.

Compared to the brainstem, the cerebellum, the backward bulge of the hindbrain, is not critical for life. But it is critical for a *normal* life. Around the time of Legallois, another French physiologist, Marie-Jean Pierre Flourens, was testing other parts of the brain in a systematic way. He proved that animals such as rabbits and pigeons could live without an intact cerebellum, but that they lost control of their movements, which became wild and exaggerated. This meant that the cerebellum had to be involved, somehow, in coordinating the muscles to help with walking, running and so on. It is involved in the acquisition of physical skills that become habitual once they have been learned.

Lower Part of the Brainstem

The first animals in possession of the equivalent of a hindbrain probably evolved more than half a billion years ago, meaning that the brainstem

A Brain of Many Parts

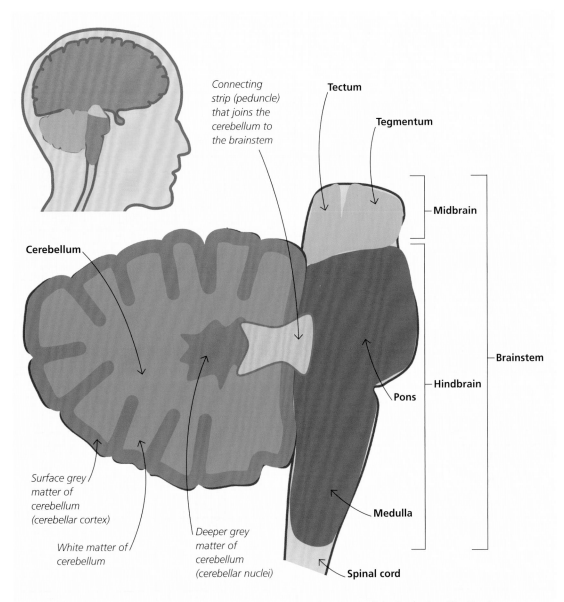

Connecting strip (peduncle) that joins the cerebellum to the brainstem

Tectum

Tegmentum

Midbrain

Cerebellum

Brainstem

Hindbrain

Pons

Surface grey matter of cerebellum (cerebellar cortex)

Medulla

White matter of cerebellum

Deeper grey matter of cerebellum (cerebellar nuclei)

Spinal cord

Together, the brainstem and cerebellum control automatic functions of the body. Specifically, the medulla oblongata, the lowest part of the brainstem, is vital for life, whereas the cerebellum is needed to properly coordinate movements.

was present in the very first backboned animals. And just as it was at the foundation of their evolution, so it is pivotal in human development too. In the human embryo, the hindbrain is the brain's root: it forms as a fold of nervous tissue running down the back, which encloses a cavity that will contain the cerebrospinal fluid and becomes the lowest of the brain's four fluid-filled ventricles. As the brain grows more, the hindbrain becomes the lowest part of the brainstem, while a bulge forming at the back turns into the cerebellum.

Standing on two legs and managing not to fall over may look simple, but it requires precise control of the muscles and constant monitoring of the body position.

At the top of the spinal cord, the central nervous system merges into the lowest part of the brainstem, the medulla oblongata, often just called the medulla, which protrudes from the base of the skull. Although only about 3 centimetres (just over 1 in) long in the adult brain, within this short section are clusters of neurons that keep us alive from minute to minute: they control heartbeat, breathing and the activity of our digestive system. The term "medulla", from the Latin for "core", was first used for the spinal cord by early anatomists, so medulla oblongata effectively means "oblong-shaped spinal cord", but by the eighteenth century it was used only for the root of the brain. The medulla contains both grey and white matter, and joins to the four lowest pairs of cranial nerves, including one that connects to

the heart. If the upper, conscious, parts of the brain fail to work, or connections with them are severed, a working medulla can keep the body alive, but the body loses consciousness: it goes into a vegetative state.

On top of the medulla, there is a bulb-shaped area of the hindbrain called the pons, from the Latin for "bridge". This connects to the next four pairs of cranial nerves and, as the name suggests, acts as a link for carrying nerve impulses between the spinal cord and the higher parts of the brain.

Cerebellum

Behind the pons, the bulge of the cerebellum is a wrinkled region of the brain in the back of the head. In most backboned animals, it is a distinct dome that is clearly visible behind the swelling of the cerebral hemispheres of the higher brain, and the two parts may be similar in size. But in humans, the higher brain, the cerebrum, is so big that it overhangs the comparatively smaller cerebellum at the rear.

The term cerebrum comes from the Latin for "brain", while cerebellum means "little brain". The comparison is appropriate. Like its bigger counterpart (the cerebrum), the cerebellum is divided into two halves, or hemispheres, and is folded and grooved. This provides a large area for the surface layer of grey matter that overlies white matter. More grey matter, in the form of nuclei, is located deeper inside. As in other parts of the brain, the grey matter is where most of the coordination functions take place. Neurons in the layered cortex of the cerebellum are among the biggest of the entire brain, and are characterized by a profuse abundance of branching nerve fibres. This helps them to process many signals at once. Unlike that of the higher cerebrum, the activity of the cerebellum is not under direct conscious control (even though it can modify conscious movement). The cerebellum is the brain's movement coordinator. This involves two main functions. Firstly, it must receive information about the body's position so it can keep the body upright. In other words, it keeps the position of the body

A Brain of Many Parts

in equilibrium with its surroundings. Secondly, the cerebellum refines the brain's control of its muscles so that movements happen smoothly and are properly coordinated. Even the apparently simplest postures and movements, such as standing and walking upright on two legs, need precise monitoring and control of equilibrium and muscles.

To do its job, the cerebellum must receive plenty of information about the body's surroundings and its position. As well as receiving impulses from the spinal nerve, it also receives them from higher centres of the brain. It relies on these centres sending sensory information they have received from sense organs around the body, including touch, vision, hearing and the relative position of all the body's

parts. The cerebellum combines this information to send appropriate signals back to the higher parts of the brain that are concerned with controlling muscles: their motor centres. These signals modify the control of body muscles according to the sensory signals that have been monitored. In this way, all these movements become smoothly coordinated. Unlike the higher brain areas, the cerebellum does not generate any movements itself; it is solely involved in refinement. If it is artificially stimulated electrically, it does not trigger any new muscle contractions. And, as the nineteenth century physiologists discovered, if the cerebellum is surgically removed or disabled, movement is still possible, even though it is uncoordinated.

A single highly branched neuron from the cortex of the cerebellum can collect a huge amount of information: the branches connect with up to 200,000 synapses. This kind of brain cell, called a Purkinje cell, is a popular subject for brain cell research because of its unusually large size.

Higher Parts of the Brain

The "higher" parts of the brain are so-called because they control higher-order functions such as conscious decision-making, emotion and personality. They are dominated by the cerebral hemispheres, which sit above core areas for receiving and processing information.

When complex animals evolved different ways of exploiting their surroundings and living their lives, it was development of the higher parts of the brain (in other words, the forebrain) that, more than any other, reflected their more sophisticated behaviour. In the first backboned animals, such as fishes, amphibians and reptiles, the higher regions were mainly concerned with processing sensory information to coordinate voluntary movements. The higher parts became bigger to accommodate more of this processing and to make the processing more effective. Conscious decision-making affected the responses that were possible, and bigger brains could store more memories to make these decisions more informed. In short, animals became cleverer, which made them better at finding food, avoiding danger and raising families.

These higher parts of the brain reached their peak of development in birds and mammals. The first mammals, which were active at night, relied heavily on smelling scents, especially in their social signalling to one another. This meant that the smell centres, known as the olfactory centres, of the higher brain became particularly large. At the same time, mammals evolved improved problem-solving skills and their cognitive skills continued to develop. In humans, the higher parts of the brain are proportionately larger than those of any other animal. While the smell centres have proportionately shrunk as the olfactory sense diminished in

The first mammals were small shrew-like creatures that were active mostly at night and had a keen sense of smell. The olfactory centres of their brains were highly developed.

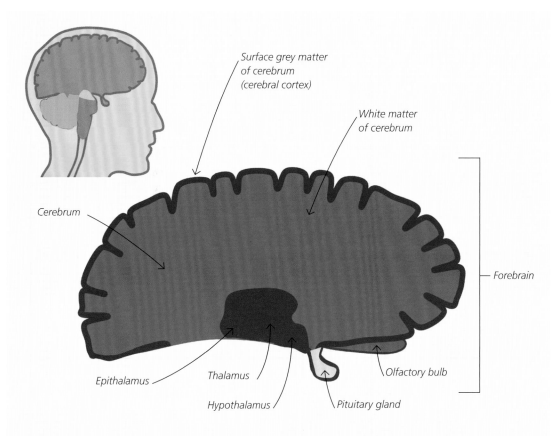

Surface grey matter
of cerebrum
(cerebral cortex)

White matter
of cerebrum

Cerebrum

Forebrain

Epithalamus

Thalamus

Hypothalamus

Pituitary gland

Olfactory bulb

The cerebrum, consisting of the two cerebral hemispheres, makes up the bulk of the "higher" forebrain. Within the cerebrum is the brain's core: a control centre composed of a paired thalamus, with epithalamus behind and a hypothalamus below. The connection with the pituitary gland, the body's "master" hormone-producing gland, reflects the importance of the brain's core in controlling the rest of the body.

importance, other capabilities, such as higher-level reasoning, emotion and language, evolved and pushed the higher brain into new territories.

Core of the Higher Brain

A sophisticated higher brain needs an impressive control hub, which is provided by its core, technically called the diencephalon. This comprises the thalamus, the hypothalamus and the epithalamus. Situated just above the brain stem, this core is the main control centre for a great many complex brain functions, including coordinating sensory information with motor responses, as well as more vital functions.

For instance, it is here that body temperature is regulated, and here that the brain communicates with the body's hormonal system. It also contains one of the brain's four fluid-filled ventricles. At its heart, this core is a relay centre. It receives sensory impulses coming up from the spinal cord, brain stem and cranial nerves. Then, after monitoring and coordinating this information, it distributes other impulses to where they are needed. Many impulses are sent up to areas of the cerebrum that control conscious activity. Signals are sent out to other parts of the body in the form of chemical hormones, to regulate functions such as the makeup of the blood or sexual development.

At the middle of the core is a part of the brain that plays a central role in the relay of the sensory information: the thalamus. It is made up of two lobes that are, anatomically, near the very centre of the brain. The thalamus is the central hub for processing sensory information and helps to give us our sense of awareness. If the thalamus is damaged, the body loses consciousness. Practically all sensory information passes through the thalamus before being relayed upwards to the sensory areas of the cerebral cortex (see illustration on previous page); only signals associated with smell pass directly to the cerebral cortex. Remarkably, some sensory pathways running through the brain seem to be duplicated. For instance, visual information takes two routes: one carries impulses directly from the eyes' retinas to the thalamus, but another must first go through

The endocrine system is made up of a number of glands around the body that release hormones. The pituitary gland in the brain coordinates the system.

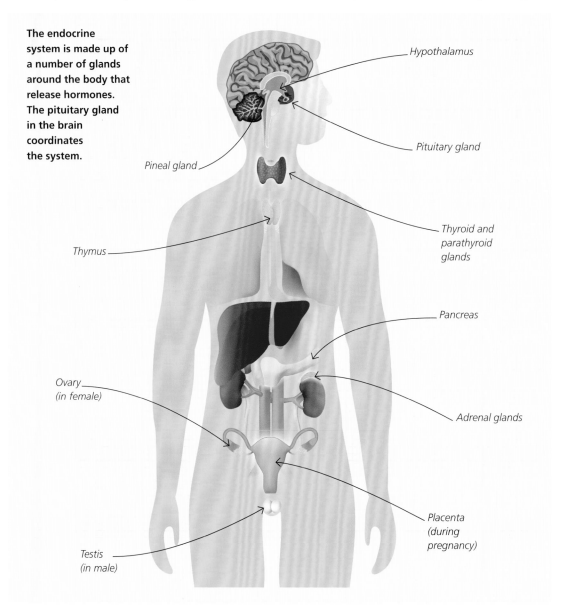

Hypothalamus

Pituitary gland

Pineal gland

Thyroid and parathyroid glands

Thymus

Pancreas

Ovary (in female)

Adrenal glands

Placenta (during pregnancy)

Testis (in male)

A Brain of Many Parts

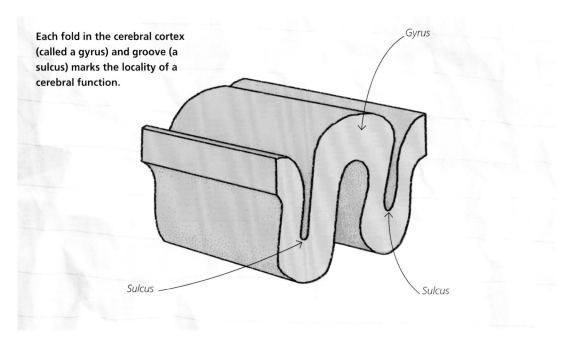

Each fold in the cerebral cortex (called a gyrus) and groove (a sulcus) marks the locality of a cerebral function.

Gyrus

Sulcus

Sulcus

the upper brainstem (specifically the tectum: see previous section). Comparing sensory information from both routes before transmitting signals to the cerebral cortex probably helps the brain to extract more information about what is being seen. The thalamus is also involved in controlling our cycles of sleep and wakefulness. When waking up from sleep, impulses associated with alertness and consciousness appear to begin in the thalamus, before passing up to the cerebral cortex.

An important role of the brain's core is to communicate with the body's hormonal system (technically called the endocrine system). Hormones are chemical messengers that are made in parts of the body called glands. These glands have a rich blood supply, helping the hormones to seep into the blood circulation. As they move around the body, they trigger specific organs to respond in a particular way. For instance, the hormone adrenalin – produced by adrenal glands above the kidneys – stimulates the heart to beat faster (among other effects). Hormones only affect specific sites because these places have special receptor proteins that bond to the hormone. The brain's core either stimulates glands to produce hormones, or produces hormones of its own. This happens

above and below the thalamus in parts called the epithalamus and hypothalamus respectively, whose effects are described in more detail in chapter 6.

Cerebral Hemispheres

The important role that the human cerebrum plays in higher brain function is reflected not just in its size but also in its structure. It makes up 85 per cent of the human brain and is folded to increase the amount of surface grey matter: the so-called cerebral cortex (cortex meaning the outer layer of an organ). A folded cerebral cortex is characteristic of most mammals, but not all of them. The cerebral surfaces of the egg-laying platypus and marsupial opossums, for instance, are smooth. The folds of the cerebral cortex are called gyri (singular gyrus), and the intervening grooves are sulci (singular sulcus). Most of these folds are clearly visible on the outer surface of the brain. Others are on the surface of the cerebral hemispheres that face inwards, inside the deep groove between the two. The arrangement of folds and grooves is always the same, as are the regions of the cerebral cortex devoted to receiving impulses from particular sense organs or regions concerned with coordination or conscious control of muscles.

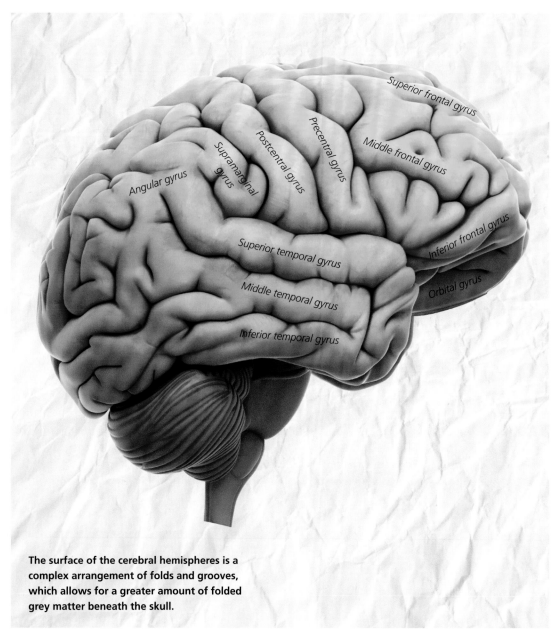

The surface of the cerebral hemispheres is a complex arrangement of folds and grooves, which allows for a greater amount of folded grey matter beneath the skull.

As with the cerebellum, the cerebrum is divided into left and right halves: the cerebral hemispheres. In general, the left hemisphere controls the right side of the body and *vice versa*. Just below each hemisphere there is an olfactory bulb, which carries sensory fibres from the nose and helps to process sensations of odour. The hemispheres are connected by bands of neurons called commissures, the most prominent of which is called the corpus callosum, from the Latin for "tough body". The corpus callosum makes up the bulkiest piece of white matter in the brain, and its abundant neuron fibres help the hemispheres to communicate with one another. The cerebral hemispheres are described in more detail in chapter 7.

A Brain of Many Parts

Chapter 6
VITAL CONTROLS

Regulating Vital Functions

The body's vital functions, such as heartbeat and breathing, are not only controlled automatically without conscious thought, but are precisely regulated according to the body's demands.

Lots of functions are important in a normal, working body, but some are more important than others. Before the body can move around, think and socialize, it must be capable of performing certain tasks that are essential just for keeping it alive. It must obtain food, grow and repair itself using its nutrients, and generate energy from food it uses as fuel. Then it must get rid of the inevitable waste products that are produced by these activities. And while doing all this, it must respond to change in an appropriate way.

As we have seen, the brain plays a pivotal role in the body's stimulus–response system. Much of this involves the kind of sophisticated decision-making that makes our behaviour so complex, and even unpredictable, when faced with a particular circumstance. But other responses are more automatic. We can be torn between choosing something from a menu, or whether to have another drink, but when it comes to boosting our heart rate in the face of a charging rhinoceros, we don't have much choice in the matter. Some simple automatic

When we are confronted with danger, such as a run-away car or a charging rhinoceros, we run, while simultaneously raising our heart rate and breathing. If a body fails to do all of these vital things, the results could be fatal.

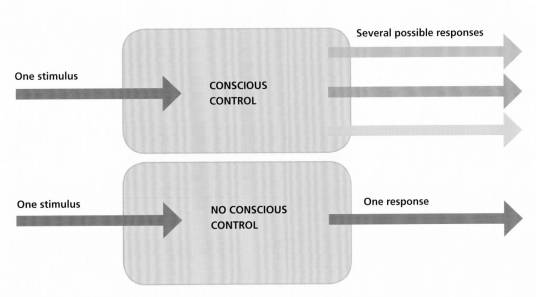

One stimulus → **CONSCIOUS CONTROL** → Several possible responses

One stimulus → **NO CONSCIOUS CONTROL** → One response

The higher parts of the brain are involved in complex decision-making processes that can respond to stimuli in lots of different ways. Control of vital functions involves more predictable stimulus–response systems.

responses such as the knee-jerk reflex (see chapter 3) come from electrical signals that shoot through the spinal cord and don't bother the brain at all. Others must use circuitry inside the brain, although we still have no obvious conscious control over them.

When to Be Automatic

Every moment of every day, our bodies are bombarded with a barrage of stimuli. Stimuli can be totally unpredictable, and the necessary responses, assuming we need to respond at all, are also not always predictable. Our "higher" conscious brains have evolved so we can decide what to do when a stereotypical automatic response is not in our interests. A lot of these stimuli come from our surroundings, such as the sound of a baby crying, a flash of lightning or the smell of freshly baked bread. Others, such as a rise in blood pressure or a sugar rush, affect sensors inside our bodies, even though an external factor might have first triggered them. It makes sense for some responses to be automatic where vital functions are at stake.

Many of these happen when internal sensors need to set off a chain of reactions that take place in the same way every time. A sugar rush after drinking a large glass of sweet lemonade must automatically trigger a series of responses that lower the blood sugar back to normal, safe levels. No conscious decision is needed, or desirable, to achieve this because there is only one outcome desired.

Even though these responses are automatic, the kinds of circuitry involved in delivering them can still be more than the simple knee-jerk reflex. They involve parts of the brain, but these are parts that are set away from others that are continuously busy with making conscious decisions. It means you don't have to think consciously about dealing with blood sugar or pumping blood every moment of every day. Some automatic parts of the system, such as breathing, can, in fact, be intercepted by conscious thought (so you can consciously decide to hold your breath, or breathe more deeply), but in the normal run of things, your conscious brain lets the automatic control system do its job.

Even when you are involved in strenuous mental or physical activity, your brain is busy regulating your temperature and heart rate.

The Need to Automatically Regulate

Some of the most important automatic systems in the body involve regulation. Many aspects of the body's operations must be constantly checked so they do not escalate or drop. These are factors that are intimately tied up with vital functions such as feeding and breathing that keep the body alive. Blood sugar, oxygen level, body temperature and heart rate must all be maintained within reasonable limits to keep cells working properly. For each one, an automatic stimulus–response system is in operation. On average, the heart of an adult human beats around 70 times a minute when the body is at rest, and breathing is somewhere between 10 and 20 breaths per minute. Automatic parts of the brain maintain the status quo. If the heart starts to go too fast, an automatic response is triggered by changes in blood chemistry to slow it back down, and *vice versa*. The circuitry needed to do this is in the brain.

Not only does the brain help to regulate, but it helps the body to adapt to circumstances and tweak the regulation accordingly. During exercise, the heart must beat faster to deliver the extra oxygenated blood. The brain monitors what the body is doing to make this happen.

How to Regulate

For each changeable factor that might affect the well-being of the body's cells, there is an ideal range: an optimum. The resting heart rate is different in each individual, but there is a rate for each person that is right to satisfy the needs of the cells. A body temperature of very close to 37°C (98°F) is just right for everyone. The control systems of the body, including the brain, are "set" to achieve these optimums, a bit like setting a thermostat for temperature control. If the level of, say, heartbeat or blood temperature, deviates away from the optimum, the automatic stimulus–response system kicks in to correct it and bring it back to the optimum. This is called negative feedback, and it keeps conditions inside the body close the optimum. It will happen even when conflicting factors in the surroundings change. Our bodies try to keep our blood at 37°C (98°F) even when it is colder outside. This internal regulation is called homeostasis.

Homeostasis succeeds by negative feedback because the body is continually and automatically detecting changes from the optimums and counteracting (negating) them. If a working heart, by chance, rises a bit above the optimum 70 beats per minute, sensors inside the body detect the rise and send an impulse to the brain's heart-control centre. The brain then sends impulses back to the heart to slow it back down. When the

slowing heart then overshoots and drops below 70 beats per minute, others sensors detect the drop and send impulses back to the brain to speed it back up. And so on. As long as the sensors are sufficiently sensitive, this system can ensure that the heart beats within a narrow range. The same process works for factors such as temperature and blood sugar level. In the next section we will see how the sensors and brain can do this.

These automatic control centres of the brain are found in the brainstem, between the spinal cord and the parts involved in "higher" conscious thoughts. As we shall see, two parts are especially important: the medulla oblongata, close to where the spinal cord joins the brain, and the hypothalamus, further up.

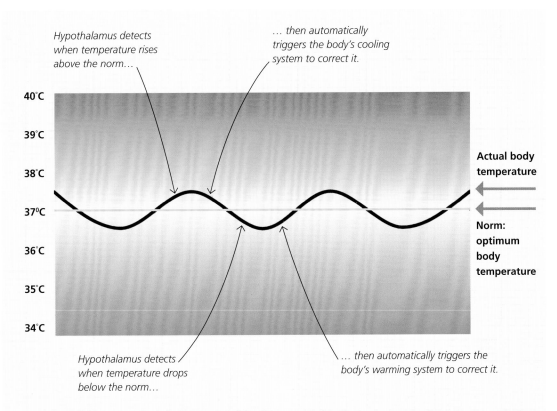

Hypothalamus detects when temperature rises above the norm…

… then automatically triggers the body's cooling system to correct it.

Actual body temperature

Norm: optimum body temperature

Hypothalamus detects when temperature drops below the norm…

… then automatically triggers the body's warming system to correct it.

The hypothalamus contains cells called thermoreceptors that continually monitor the temperature of the blood and activate appropriate responses to keep the blood temperature close to the norm. In addition, the hypothalamus receives information from thermoreceptors in the skin. The medulla oblongata is also involved in the process of automatic regulation of other factors.

Nerves versus Hormones

Nerve impulses are the fastest way for the brain to regulate specific organs. Bioelectric impulses passing through the nerves and the brain move with lightning speed. Nerves connect to precise points, meaning that they can target very specific places. The impulses from the heart-control system in the brain only target the heart because the nerves involved connect to a patch of heart muscle that is responsible for provoking the cycle of contractions involved in the heartbeat. The chemical neurotransmitter that provokes the response is released at the end of the nerve (see chapter 2). But it spreads so little from its source (keeping within the synapses) that its effect is localized. Other chemical signals in the body are released into the blood circulation and have more far-reaching effects. These chemicals are the hormones.

Hormones are made in particular glands (called endocrine glands) that are dotted around the body. Unlike the "surface" glands that discharge their contents, such as sweat and oil, over the skin or from the gut wall into our digestive system, hormone-producing glands pass their products into the blood. This means they have no obvious opening (duct). Instead, their hormone-making cells are clustered around tiny blood vessels. Hormones work like chemical signals, just like the neurotransmitters coming from neurons. Like neurotransmitters, hormones trigger responses in target cells by latching on to special

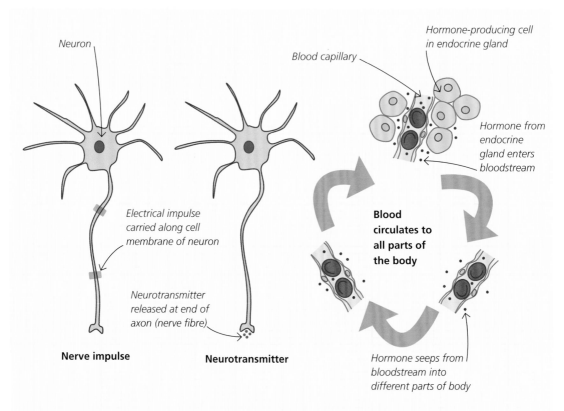

Neuron

Electrical impulse carried along cell membrane of neuron

Neurotransmitter released at end of axon (nerve fibre)

Nerve impulse

Neurotransmitter

Blood capillary

Hormone-producing cell in endocrine gland

Hormone from endocrine gland enters bloodstream

Blood circulates to all parts of the body

Hormone seeps from bloodstream into different parts of body

Neurons are involved in two kinds of signalling: electrical impulses move rapidly down long nerve fibres, then trigger the local release of chemical neurotransmitters at synapses. This carries a fast signal long-distance to a very specific destination. Hormones move more slowly through the bloodstream, but can trigger responses all around the body.

Vital Controls

The hormone adrenaline is released by the body in stressful situations, such as the start of a race. It prepares us for "fight or flight" by increasing the heart rate to pump more blood.

receptors in their cell membranes. However, hormones can spread much further around the body than neurotransmitters. Hormones can reach just as far as the impulses that travel through nerves, but the blood circulation allows them to spread widely. Their effects are slower and more long-lasting. For instance, a tiny amount of hormone from the adrenal glands above the kidneys would be detected in a blood sample taken from the finger, and might still be present days later.

When to Signal with Hormones

The wide-ranging spread of hormones is matched by their wide-ranging effects, even though they are usually involved in achieving some common goal. For instance, the hormone adrenaline (from the adrenal glands) not only makes the heart speed up, but also dilates the pupils. This happens because

both these parts of the body have the necessary receptor, so both will respond to the adrenaline. And the two effects are useful under the same circumstance, such as being excited, fearful or shocked, so it pays for the same hormone to stimulate the related responses.

Hormones, then, are often involved in provoking more long-lasting responses that have to involve a coordinated impact on many body parts at the same time. Unsurprisingly, this means they are closely associated with general states and moods, and states that can be very persistent indeed. Hormones are the chemical signals that are responsible for controlling aspects of development that last years (see next section). From the point that we produce our first hormones inside the womb, hormones help to make our bodies male or female. And, years later, during puberty, they are still at work getting us ready for sexual maturity.

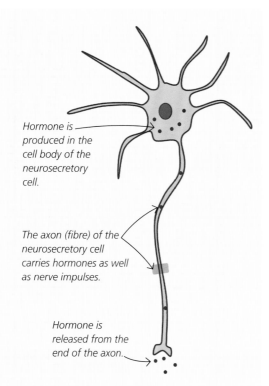

Hormone is produced in the cell body of the neurosecretory cell.

The axon (fibre) of the neurosecretory cell carries hormones as well as nerve impulses.

Hormone is released from the end of the axon.

Some neurons in the hypothalamus can produce hormones of their own. These neurosecretory cells are where the nervous system and hormonal system come together.

Hormones and the Brain

Of course, the brain is not detached from our hormones. In fact, the brain contains many of the receptors that pick up hormones, so it is sensitive to them, just like many other parts of the body. This means that moment-to-moment brain activity is affected by these chemical signals, and this can influence some of our "higher" brain functions, such as mood and emotion. It helps to explain, for instance, why the sex hormones that control our reproductive systems during development can wreak so much havoc on our moods during puberty and beyond.

But the brain is not just a responder in the hormone system. It can also produce hormones of its own and is vital in the way other hormones are

controlled. In the 1940s, German-born neuroscientists Ernst and Berta Scharrer were the first to observe droplets containing hormones in microscopic views of brain cells. They were looking at the hypothalamus, the part of the brain's core already known to be involved in the automatic regulations that form the basis of homeostasis. The hypothalamus has a special relationship with the body's hormonal system: it sits just above the "master" gland of this system, the pituitary gland, and connects to it via a short stalk. The hypothalamus had long been known to communicate with the pituitary, but it took the Scharrers to demonstrate that it did so not only by electrical impulses, but also by its own hormones.

How the Brain Communicates with the Hormone System

Among other things, the hypothalamus is the brain's control hub in coordinating the activities of the nervous and hormonal systems: it is here that the two physically come together and communicate. On one side, the hypothalamus is linked to the rest of the brain by its conventional neurons; on the other, it "talks" to the hormonal system by controlling its pituitary master. For instance, the hypothalamus may send nerve impulses to the pituitary to make it release hormone A. When this hormone circulates around the body, it might then affect another gland to release hormone B, and so on. Overall this means that the brain, including the hypothalamus and everything it is connected to, can influence hormones, and hormones can in turn influence the brain.

The hypothalamus cells that release their own hormones can also carry nerve impulses. Because of their dual function, they are described as neurosecretory. (Hormones are said to be *secreted* into circulation.) As we shall see (later in this chapter), whether the pituitary is triggered by impulses or brain hormones depends on the kinds of hormones involved: the stalk between the hypothalamus and pituitary is packed with neuron circuits and blood vessels, meaning that it carries both kinds of signal.

The Brain and the Immune System

The influence of the subconscious brain in the day-to-day running of the body clearly reaches far beyond the routine check of body temperature, heartbeat, etc. Through its physical link with glands, our behaviour can be altered by mood-changing hormones. But can our brain be affected by our physical wellbeing: by the immune system that fights disease? In times of stress, for instance, we may be more vulnerable to infection. Does the brain communicate with the immune system like it communicates with the hormonal system? Until recently, the brain was thought to be largely separate from the white cells involved in immune reactions. The blood–brain barrier protected the brain from the onslaught of a white cell attack: the brain was said to be "immunologically privileged". Instead, it was protected by its own army of infection-fighting glial cells. However, recent discoveries have shown that the brain has lymphatic vessels. These are drainage vessels that link up with nodes that carry white cells. They pervade the body, but were previously thought to be missing from the brain. The newly found brain lymphatics are being described as the "missing link" between the brain and the immune system, and may help improve understanding of diseases associated with degeneration or inflammation of the brain, such as Alzheimer's disease or multiple sclerosis.

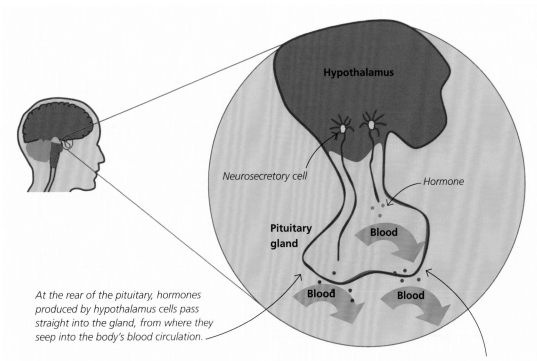

Hypothalamus

Neurosecretory cell

Hormone

Pituitary gland

Blood

Blood

Blood

At the rear of the pituitary, hormones produced by hypothalamus cells pass straight into the gland, from where they seep into the body's blood circulation.

Many kinds of hormone enter the general blood circulation from the pituitary gland beneath the brain. Some of these have been produced by the brain's hypothalamus. Others are made in the pituitary gland, but are released when prompted to do so by other hormones coming from the hypothalamus.

At the front of the pituitary, hormones produced by the hypothalamus flow down to the gland in blood, where they trigger other kinds of hormones to seep into the body's blood circulation.

The Autonomic Nervous System

When the central nervous system affects the actions of muscles and glands over which we have no conscious control, its electrical impulses pass through nerves that make up the autonomic nervous system.

The earliest anatomists knew that the brain was somehow involved in controlling vital functions because they saw that there were nerves leading from the brain to the vital organs. Centuries of experiments involved cutting these nerves and observing the results. This information brought physiologists closer to understanding exactly what each nerve did, but it wasn't until the late 1800s that the British physiologist John Newport Langley brought this so-called "vegetative" nervous system into finer focus. He coined a new name for it: the autonomic nervous system, and studied it by using techniques that involved interrupting synapses with signal-blocking chemicals such as nicotine (see chapter 2). In this way, Langley was able to show what the autonomic nerves were doing to particular organs. It was a complicated picture: some nerves released neurotransmitters that excited certain organs but slowed others down.

The autonomic nervous system connects not only to the heart and lungs, but also to glands around the body, as well as all the sensors needed to pick up the information that is needed for regulation. The part of the brain that is its overall control centre is right at the base of the brainstem: the medulla oblongata.

The rest of the nervous system, the section under conscious control, is called the somatic nervous system. Some of its nerves are connected to muscles that move the skeleton: the ones that we can control at will. Other nerves of the somatic system send impulses from sense organs to the conscious parts of the brain.

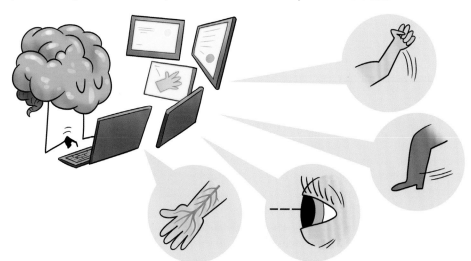

The medulla carries on controlling our heartbeat and breathing, with no interference from the conscious brain, helping to ensure that the blood keeps moving and the lungs keep it well-oxygenated. Without oxygenated blood, cells around the body quickly die. You don't get any more vital than that.

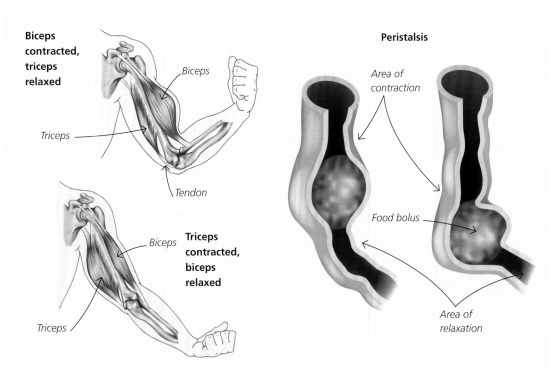

Biceps contracted, triceps relaxed

Biceps

Triceps

Tendon

Biceps

Triceps contracted, biceps relaxed

Triceps

Peristalsis

Area of contraction

Food bolus

Area of relaxation

When we consciously move a part of the body, such as the arm (left), impulses from the conscious brain (the cerebrum) are sent to skeletal muscles to pull on bones. Movement of the soft organs, such as peristalsis in the gut (right), involves impulses from our unconscious brain (the medulla oblongata).

Voluntary and Involuntary Muscles

The most familiar muscle groups in our body come in pairs that have opposite effects: they are said to be antagonistic. We need a biceps muscle to flex the arm, but another one, the triceps, to straighten it. The two muscles have opposite effects because they are connected to the skeleton in different ways. Muscle pairs like this are necessary because a muscle can only pull; it cannot push. When a stimulated muscle contracts, it becomes shorter and pulls two bones together around a joint. To get the joint to return to its original position, an antagonistic muscle is needed to pull it back again. These actions are conscious: they happen under the influence of the somatic nervous system, with the conscious brain having overall control. Separate motor nerves are needed to go to the two antagonistic muscles.

Muscles of the soft organs are not like this. These muscles are physically different: skeletal muscles look striped under the microscope, whereas the visceral muscles associated with soft organs are stripeless and described as "smooth". They also work in a different way. Muscles of organs such as the heart or gut are packed inside the organ walls, and when they contract, they usually squeeze down, moving whatever is inside. Heart muscle moves blood; gut muscle moves digested food. When their muscles relax, the organs return to their dilated state, so no opposite antagonist is needed. But organs work continuously and rhythmically, and the autonomic nervous system must be able not only to trigger a response in the muscles, but also to change the rate of activity. The medulla, for instance, controls the rate of heartbeat, breathing and the rhythmic squeezing action in the gut that moves food through it.

Fight or Flight or Rest and Digest?

Although our emotions and moods vary in many different ways, as far as the vital organs are concerned there are only two states that really matter: inaction and action. When the body is motionless, its skeletal muscles are resting. (In fact, even in a motionless body, whether standing or sitting, there are always some muscles that are contracted to maintain posture, but overall the demands on the musculature are low.) But when the body starts moving about, these muscles contract more to move the skeleton. And this needs more energy, which is supplied from more respiration: the chemical process that releases energy from reactions that consume glucose and oxygen. The extra glucose and oxygen are provided by boosting the blood supply, which means raising the heartbeat, as well as breathing faster to get more oxygen into the circulation. If the body moves as much as it can, and reaches peak level of exercise, the heartbeat and breathing also peak.

The automatic control systems centred on the brain are responsible for making the organs ready for action. And, as usual, sensors in the body detect the necessary stimuli. If you run from a resting state, your muscles start working straight away, so chemical respiration goes faster. It uses more glucose and oxygen, but also generates more waste products, such as carbon dioxide and lactic acid. Any of these chemical changes might be picked up by the sensors and can get

Even though these runners are tired, they consciously control their leg muscles to keep running. Meanwhile, the muscles controlling breathing unconsciously keep their lungs working at peak levels.

Vital Controls

During physical activity, the medulla oblongata sends signals to the gut to slow down digestion. The medulla switches to boost the gut and speed digestion when the body is at rest.

communicated to the brain to automatically speed up the heart and lungs. As a result, the body is alert and active. But not all organs work more quickly during exercise. The gut functions best when the body is resting. With less blood needed for the skeletal muscles, more of it can be diverted to the digestive system to help work its walls and absorb more nourishment.

The contrasting effects of action and inaction are controlled by different parts of the autonomic nervous system. John Newport Langley, the originator of the idea of the autonomic nervous system, recognized this and came up with different names for the different parts. He called the part that prepares the body for physical activity the sympathetic nervous system. The name came

from an antiquated idea of "sympathy": the vague notion that organs worked together in coordination. But Langley reinvented the term for more precise usage. He called the opposite part, the one associated with the resting state, the parasympathetic nervous system. Both autonomic systems have their control centres in the medulla oblongata, but they rely on different nuclei of brain cells, involve different nerves and use different neurotransmitters. The sympathetic nervous system gets the body into "fight or flight" mode: it speeds up the heart and lungs, but slows the gut down. Then, during rest, it is the parasympathetic nervous system that is firing impulses to slow down the heart and lungs and gets the gut churning instead.

If the heart is beating too slowly…

1. *blood accumulates here and stretches the veins…*

2… *which sends impulses along a sensory nerve to the medulla…*

3… *which sends impulses along a sympathetic motor nerve back to the heart to speed it up.*

This circuit passes through the medulla's cardioaccelerator centre.

If the heart is beating too fast…

1. *blood accumulates here and stretches the arteries…*

2… *which sends impulses along a different sensory nerve to the medulla…*

3… *which sends impulses along a parasympathetic motor nerve back to the heart to slow it down.*

This circuit passes through the medulla's cardioinhibitory centre.

Automatic regulation of heartbeat by the brain's medulla stops the heart beating too slowly or too fast and involves both sympathetic and parasympathetic nerves – to speed it up or slow it down. Sensory nerves are shown in blue; motor nerves in orange. Other factors, such as blood pressure and blood acidity, will affect the heart's speed too.

How the Brain Controls the Heart Rate

The mechanism by which the brain controls heartbeat is a good example of how the dual parts of the autonomic nervous system work. The medulla contains nuclei of cells in its cardiovascular centre. As we have seen, one group of cells is concerned with speeding the heart up, while another slows the heart down. They communicate with the heart through sensory and motor nerves.

If the heart is beating too slowly, blood tends to build up in the veins leading into the heart, making them stretch. Sensors in the walls of the veins are stimulated and send impulses to the medulla along a sensory nerve. Specifically, this nerve targets the part of the cardiovascular centre concerned with speeding the heart up. The centre does this by sending an impulse back to the heart along a motor nerve. This time, the heart might beat too fast, so blood now accumulates in the main artery leading away from the heart, making it stretch. The artery's sensors send impulses up to the medulla, targeting the part concerned with slowing the heart down. This sequence is repeated in a cycle

of self-regulation. The heart's muscle can respond differently to the two kinds of motor nerves because they release different kinds of neurotransmitter (the chemical signal between synapses: see chapter 2). One speeds it up, the other slows it down.

The motor nerves from the medulla in this self-regulation system provoke a part of the heart wall that is the organ's natural pacemaker: the medulla sets the overall speed at 70 beat per minute, or thereabouts. But the medulla is also influenced by exercise-related factors that are detected by other sensors in the arteries leading to the neck. For instance, if the body runs, waste products from the skeletal muscles make the blood slightly acidic, which speeds the heart up.

The heart muscle is unique in that it can stimulate itself to contract. All other muscles, including skeletal muscles and those used for breathing and gut movement, will only contract with impulses from a motor nerve. By contrast, heart muscle generates its own impulses. If the nerves connecting to the heart are severed, it keeps beating, although it will do so erratically. Outside impulses are needed to regulate the beat.

How the Brain Controls Breathing and Digestion

A similar kind of reflex control happens when we breathe, both to ensure that we breathe in and out in a regular cycle and to keep the breathing rate in check. The muscles involved include the diaphragm in the floor of the chest cavity and those between the ribs. Collectively, when these muscles contract, the chest expands and we breathe in. This time the "breathing centre" in the medulla is involved: one part triggers breathing in, while the other triggers breathing out. As air fills the lungs, the walls of the airways stretch, stimulating sensors. They send impulses back to the medulla to switch off the "breathing in" centre and activate the "breathing out" centre. Just like the heart, the medulla is also affected by waste products in the blood. If carbon dioxide builds up, for instance, it makes the breathing centre work faster, and our breathing rate increases to get rid of the build-up. But unlike the heart, the lungs are also under conscious control from the higher parts of the brain. This means that you can consciously breathe faster or more slowly, or even hold your breath.

Automatic control of the gut depends on lots of different sensors, some of which work even before food hits the stomach. Sight, smell and even the sound of cooking can provoke the brain's gut centres, which are found in the medulla and hypothalamus. And when food passes through the digestive system, a combination of stretched gut walls and the presence of chemical nutrients will make the brain keep activating the gut wall muscles, as well as prompting the release of digestive juices. In fact, a special centre in the medulla is also involved in getting food to the stomach in the first place, by controlling the swallow reflex.

Dealing with Irritants

The body takes advantage of the brain's automatic controls to get rid of anything irritating: something that starts with an irritation can end up being dangerous. Microbes and poisons can get into the body through the mouth and nose, so there has to be a way of getting them out. Sneezing, coughing and vomiting are ways of dealing with irritants by forcibly expelling them from the body. In each case, the medulla triggers the reflex in response to the irritant.

When we breathe in (inspire)…

1. the diaphragm muscle pulls down, which pulls air into the lungs and stretches the airways.

2. This sends impulses along a sensory nerve to the medulla.

4. This then stops the diaphragm from contracting, so we automatically breathe out.

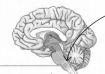

3. The impulses enter the medulla's expiratory centre and simultaneously suppress its inspiratory centre.

When we breathe out (expire)…

1. the diaphragm muscle relaxes and springs back up, forcing air out and relieving the stretching.

2… so no sensory impulses are sent to the medulla.

3. This switches the inspiratory centre back on…

4… so impulses are sent along a motor nerve to the diaphragm, stimulating it to contract so we breathe in again.

The brain's medulla ensures that breathing in is automatically followed by breathing out because the two centres involved inhibit one another. The inhibition takes two or three seconds, during which our diaphragm switches from contracting to relaxing. More carbon dioxide in the blood reduces the switch-over time, making us breathe more rapidly.

Chemical Signalling

Activities in the brain are closely linked to chemical signals such as neurotransmitters and hormones. These chemicals are involved in controlling many aspects of the body's behaviour, its development and even mood.

At any one moment, a complex cocktail of chemicals is provoking a multitude of actions inside the body, and many have a big influence on the brain. We have seen how neurons produce chemicals called neurotransmitters that help them to communicate among themselves and with body parts such as muscles. Although these only work over tiny distances, different varieties of neurotransmitters might have completely opposing effects: one could speed the heart up, another might slow it down. Other chemical signals also work over small distances, but in places might spread to affect more cells. Histamine is a neurotransmitter in the central nervous system, but elsewhere it makes tissues become inflamed to help microbe-eating white cells get to a site of possible infection. Hormones travel even further in the blood circulation and can affect lots of different places at once.

Deep inside the brain, neurotransmitters can help to swing a person's mood from joy to despair within a matter of seconds.

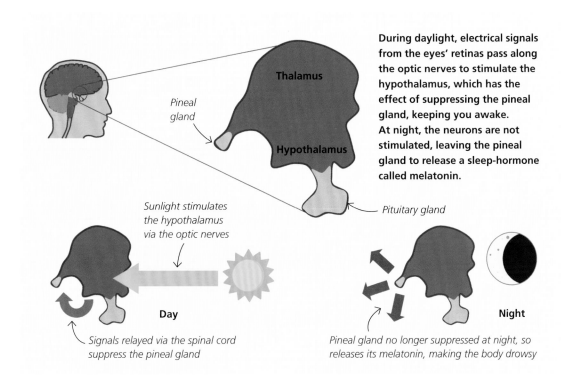

Thalamus

Pineal gland

Hypothalamus

During daylight, electrical signals from the eyes' retinas pass along the optic nerves to stimulate the hypothalamus, which has the effect of suppressing the pineal gland, keeping you awake. At night, the neurons are not stimulated, leaving the pineal gland to release a sleep-hormone called melatonin.

Pituitary gland

Sunlight stimulates the hypothalamus via the optic nerves

Day

Signals relayed via the spinal cord suppress the pineal gland

Night

Pineal gland no longer suppressed at night, so releases its melatonin, making the body drowsy

As we have seen, the brain's connection with the pituitary gland (the master hormone-producing gland) beneath the hypothalamus is an important means of coordinating hormonal responses of the body with what is going on in the brain. Although it is only the size of a pea, the pituitary gland is responsible for controlling more bodily processes than any other gland. But the pituitary is not alone. A short distance further up and further back, near the top of the brain's core, is another gland that is even smaller. The pineal gland is scarcely bigger than a grain of rice, but it is important in something that we often take for granted: our innate sense of the passage of time.

The Brain's Internal Clock

Like the pituitary, the pineal gland is attached to the brain's core. But while the pituitary lies under the lowest part of the core, the hypothalamus, the pineal gland is deeper inside, forming part of the core called the epithalamus. The pineal gland is thought to have evolved

from a light-sensitive organ. In fishes, amphibians and reptiles, it detects light, and in one reptile, the tuatara, it even functions as a rudimentary "third eye". In humans, its link with light is more subtle. It helps to detect the cyclic rhythm of day and night, and is affected by signals from the retinas of the eyes. These impulses are picked up first by nuclei of neurons in the hypothalamus. These set the body's internal clock, configured by the 24-hour cycle of daylight and darkness. During darkness, the pineal gland releases its hormone, melatonin. Like other hormones, melatonin then binds to specific receptors, most of which are in the brain. Its effect is to make the body sleepy. During daylight, the internal clock suppresses the pineal gland, so melatonin is not released. (Interestingly, in nocturnal animals melatonin has the reverse effect: it keeps the animals awake.) This means that, as levels of melatonin rise and fall as day follows night, it controls the cycle of sleep and wakefulness. Relaxation techniques such as yoga and meditation have been shown to increase the body's melatonin levels.

Controlling Sexual Development

Human sex is determined by genes and chromosomes (see chapter 11), which control the levels of hormones that affect sexual development. The two main sex hormones, oestrogen and testosterone, are produced by the sex organs: more oestrogen comes from ovaries, and more testosterone from the testes, and it is the balance between the two that determines sexual differences at puberty. In fact, these hormones trigger responses in many parts of the body, and both act on the brain to affect behaviour; both seem also to be needed for a healthy sex drive. But although oestrogen and testosterone are chemically similar, they affect the brain in very different ways. Oestrogen, for instance, may make the brain more susceptible to the effects of stress, explaining why women are twice as likely to suffer from stress-related illnesses such as depression. Testosterone, in general, increases aggressive behaviour, but testosterone levels in men diminish during fatherhood, or when they fall in love.

Other hormones that influence sexual development are produced by the pituitary gland under the influence of the brain. These so-called gonadotropins come from the front part of the pituitary, where they are released in response to other chemical triggers produced in the

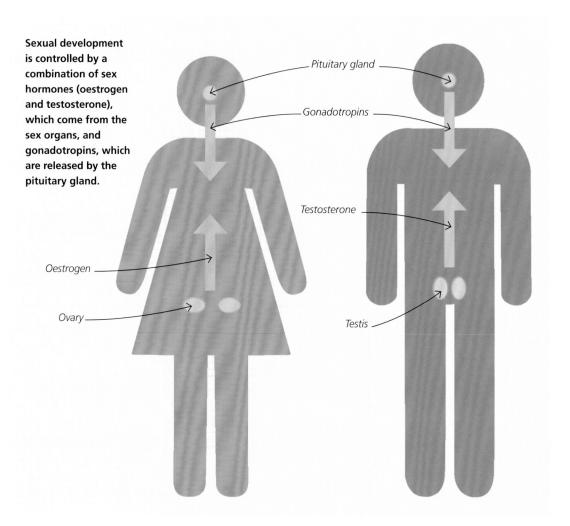

Sexual development is controlled by a combination of sex hormones (oestrogen and testosterone), which come from the sex organs, and gonadotropins, which are released by the pituitary gland.

Pituitary gland

Gonadotropins

Testosterone

Oestrogen

Ovary

Testis

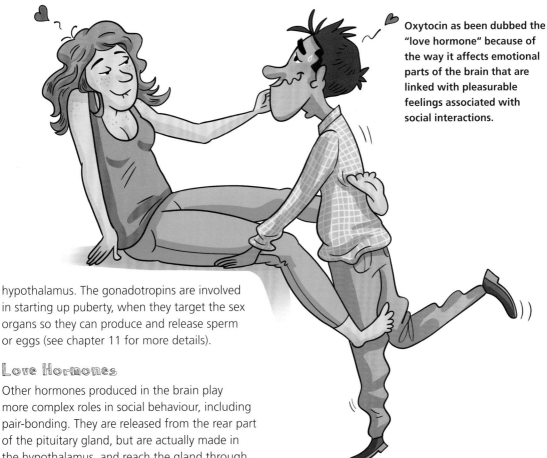

Oxytocin as been dubbed the "love hormone" because of the way it affects emotional parts of the brain that are linked with pleasurable feelings associated with social interactions.

hypothalamus. The gonadotropins are involved in starting up puberty, when they target the sex organs so they can produce and release sperm or eggs (see chapter 11 for more details).

Love Hormones

Other hormones produced in the brain play more complex roles in social behaviour, including pair-bonding. They are released from the rear part of the pituitary gland, but are actually made in the hypothalamus, and reach the gland through neurosecretory cells. One of these hormones, vasopressin, is better known for its effect on the kidneys: it stimulates them to retain more water, so less is lost in urine. Vasopressin levels therefore rise when the body is dehydrated and needs to conserve water, such as after strenuous exercise. But vasopressin has other effects: for some, at least, it encourages friendliness.

Vasopressin belongs to a family of hormones that are intimately linked to aspects of social behaviour. A related hormone from the hypothalamus–pituitary system, called oxytocin, has even closer ties to sociality that begin in the womb. When a baby reaches full term, it makes the womb more sensitive to oxytocin, which stimulates the womb's muscles to contract, bringing on labour. Oxytocin continues to play a role during nursing: it stimulates breasts to release milk in the so-called "let-down reflex". As the baby sucks on the nipple, nerve impulses are sent from the breast back to the hypothalamus, which responds by releasing surges of oxytocin to trigger the "let-down" of milk from mammary glands. In this way, oxytocin helps to secure a greater bond between mother and child. But its effects go far beyond even this. Not only is oxytocin involved in maternal love, but also in romantic attachment, infatuation, trust, empathy and even orgasm. Its effects seem to be more marked in women than men, and testosterone has been shown to depress oxytocin levels. Abnormally low levels of oxytocin have been observed in people with autism, possibly helping to explain their problems with social interactions.

Stress Responses

The brain has a remarkable capacity for responding to stressful situations when there is a perceived level of threat. In animals, a stressful situation might arise because there is imminent danger or lack of food. The body must take appropriate action to reduce or remove the threat, and so enters an aroused state to make this happen as quickly as possible. Stress involves the sympathetic part of the autonomic nervous system (see previous section), the system that prepares the body for "flight or fight". This might have several effects, such as increasing heart rate, raising blood pressure or releasing more stored sugar to provide energy-releasing fuel. These and other factors also have important impacts on the brain. In the absence of any genuine need for "fight or flight", its persistent effects on the brain can lead to illness.

When events signal a stressful situation to the brain, it initiates a chain reaction of hormones. First, the amygdala stimulates the hypothalamus to produce hormones that target the front of the pituitary gland, just like the ones that prompt the release of gonadotropins. The pituitary gland, in response, releases a hormone called ACTH (adrenocorticotropic hormone). This, in turn, makes adrenal glands above the kidneys release yet another hormone called cortisol.

The Brain's Neurotransmitters

Sensations and mood, such as feelings of love or stress, depend not only on hormones circulating

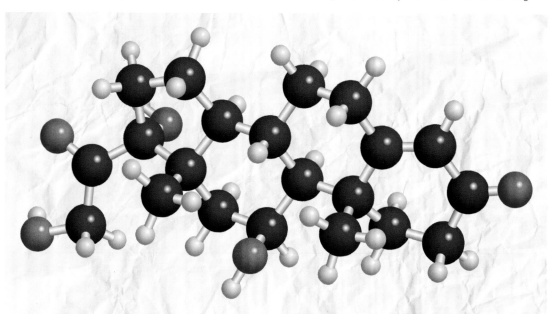

The Stress Hormone

Cortisol is the body's "stress hormone". One molecule of cortisol is shown above, combining 21 carbon atoms (black) with 30 hydrogen atoms (white) and five oxygen atoms (red). This chemical has a big impact on the higher workings of the brain. As well as preparing the brain for fight or flight, cortisol makes the brain hold on to emotive memories: memories of the sorts of things that might have produced the stress. This sounds counterproductive, but it serves a purpose. By making the brain fixate on stress-causing events, it emphasizes the need to avoid them in the future. Unfortunately, the effects of stress can linger. If cortisol levels stay high, this can cause unpleasant side effects. It can diminish the immune response, making the body more susceptible to infection and disease, or aggravate the brain, leading to anxiety or depression.

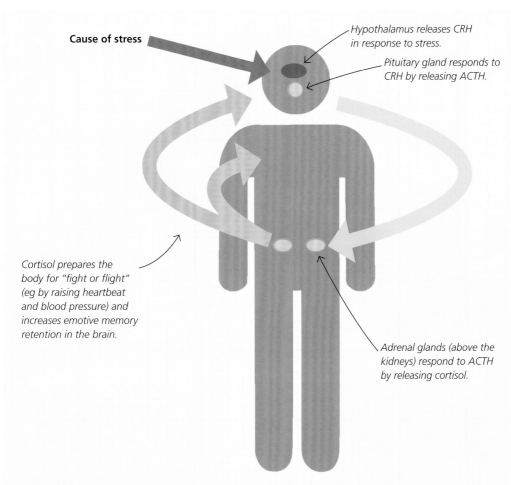

Cause of stress

Hypothalamus releases CRH in response to stress.

Pituitary gland responds to CRH by releasing ACTH.

Cortisol prepares the body for "fight or flight" (eg by raising heartbeat and blood pressure) and increases emotive memory retention in the brain.

Adrenal glands (above the kidneys) respond to ACTH by releasing cortisol.

The hypothalamus responds to stress by releasing CRH (corticotropin-releasing hormone), which then stimulates the pituitary gland to release ACTH (adrenocorticotropic hormone). ACTH then triggers the adrenal glands to release cortisol, the hormone that reacts with the body to give the physical symptoms of stress.

around the body in the bloodstream, but also on chemical signals that have more local effects in the brain. These are the neurotransmitters that cross synapses, the gaps between nerve fibres of adjacent neurons (see chapter 2). There are probably over 100 different kinds of neurotransmitter at work in the central nervous system, but only about a dozen or so do most of the work.

The most dominant neurotransmitter that excites neurons in the brain is called glutamate, which is an amino acid: one of the building blocks for making protein. A related substance, called GABA (gamma-aminobutyric acid) has an opposite effect: it inhibits neurons. Together, glutamate and GABA account for around 90 per cent of the neurotransmitter activity in the brain. Their effects are so pervasive that, when cells' synapses are regularly triggered, some of them strengthen while others weaken and disintegrate. As a result, synaptic connections can change over time with use, a process that forms the physical basis for storing memories (see chapter 8).

Eating delicious food is linked to the release of serotonin, leading to feelings of pleasure.

Some brain neurotransmitters are produced more locally than glutamate and GABA, but they can still have wide-ranging effects. Serotonin is produced in nuclei of the brainstem (specifically in the medulla and midbrain) and is associated with feelings of happiness; low serotonin levels have been found in people with depression or suicidal thoughts. It has wider effects on sleep, memory and appetite. A neurotransmitter called dopamine works in a similar way. It is produced by the floor of the midbrain (a region called the tegmentum: see chapter 5) and plays a central role in the brain's

"reward" system, giving feelings of pleasure and euphoria. Both dopamine and serotonin seep into the higher parts of the brain where they are released together as rewards for performing tasks that, as far as the body is concerned, deserve to be repeated. The pleasurable feelings associated with sugary food or sex comes from these kinds of chemicals. Drugs that act either as depressants or as anti-depressants often work by targeting these reward systems. This is explored in more detail in chapter 9.

Vital Controls

Chapter 7
PROCESSING INFORMATION

Sensing the World Around Us

Sense organs contain specialized cells that are activated by stimuli. Nearly all the signals generated by stimuli are passed to the brain, which integrates the information and determines any responses that might be necessary.

The body is armed with batteries of sensors, both inside and out. Our eyes, ears, nose, taste buds and skin help us pick up on changes in our surroundings, while sensors inside our bodies continually monitor things such as blood pressure and sugar levels. Internal changes are detected and acted upon automatically through reflex actions that work through parts of the brainstem (see chapter 6), but external stimuli typically provoke responses that are based on higher-order processing. This means the signal pathways involved must pass through parts of the higher brain: the cerebral hemispheres.

Sense organs of the head communicate directly with the brain through cranial nerves, but when the rest of the body is stimulated, impulses pass through spinal nerves to the spinal cord, before passing up to the brain. All these sensory impulses, with the exception of those generated by smell, are relayed to brain's main sensory coordination hub, the thalamus, before being sent to the cerebral cortex, where they are consciously perceived. The signals from the nose pass into the higher brain through a special route: the olfactory bulb.

What Are Senses?

Sensing and perceiving are not the same thing. Our sense organs collect information about the world around us and convert this information into a form that the nervous system can send to the brain: as nerve impulses. The brain interprets this pattern of impulses (their rate, where they have come from)

A number of different sense organs in our heads respond to a variety of stimuli, including light, sounds, smells and tastes. The sense organs send information about these stimuli to the brain to be processed.

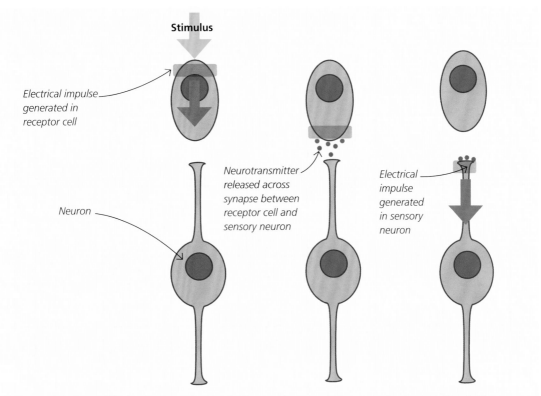

Stimulus

Electrical impulse generated in receptor cell

Neuron

Neurotransmitter released across synapse between receptor cell and sensory neuron

Electrical impulse generated in sensory neuron

The junction between a receptor cell and a sensory neuron is marked by a synapse. Just like the synapses between two neurons, a chemical neurotransmitter helps the signal to cross the gap.

and organizes the information to give us our perception of what the world is like. For instance, the eyes collect information about light: its intensity, wavelength and origin. All these factors determine the way electrical impulses are fired through the optic nerve to the brain, so we perceive what we see with shape, colour, depth, movement and so on.

Sensing begins with a stimulus. This is any sort of change in the surroundings that can make the body respond. But it doesn't provide energy for the response. If a person falls off a chair because someone else pushed them, they do so because the person doing the pushing transferred energy in the push. This is not a biological stimulus–response system. But if someone falls off because they were startled by a shout, that involves a biological stimulus. Energy comes from the fuel in the muscles, not from the person shouting. The

muscles use this fuel to contract, but the trigger to do so was the electrical impulses reaching the muscles through the nervous system.

To detect any stimulus, we need special kinds of neurons called receptors. (These should not be confused with receptor molecules that intercept chemical signals, such as neurotransmitters and hormones, at synapses.) Like other neurons, receptor cells carry changeable patterns of electrical charge on their membranes. But while other neurons are stimulated by neurons, receptors start the ball rolling, so to speak. A stimulus from the body's surroundings charges their cell membranes up with an electrical charge. Receptors in different sense organs are adapted to pick up different kinds of stimuli, but once charged up, they all pass their impulses into the nervous system.

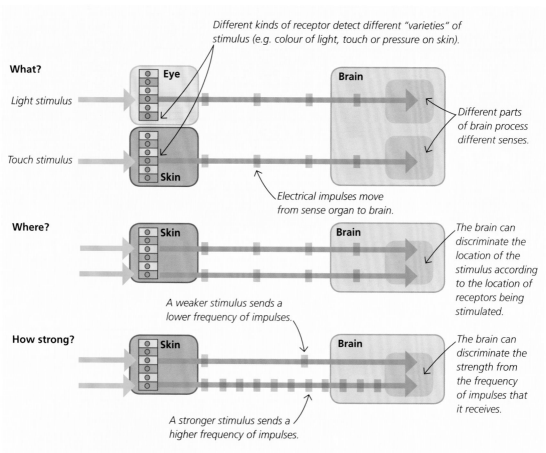

Different kinds of receptor detect different "varieties" of stimulus (e.g. colour of light, touch or pressure on skin).

What?

Light stimulus

Eye

Brain

Touch stimulus

Skin

Different parts of brain process different senses.

Electrical impulses move from sense organ to brain.

Where?

Skin

Brain

The brain can discriminate the location of the stimulus according to the location of receptors being stimulated.

A weaker stimulus sends a lower frequency of impulses.

How strong?

Skin

Brain

The brain can discriminate the strength from the frequency of impulses that it receives.

A stronger stimulus sends a higher frequency of impulses.

The brain uses the pattern of impulses arriving along sensory neurons to discriminate the kinds of sensory information being detected, to determine our overall perception of the world around us.

Discriminating Senses

The reason that the perception you experience when listening to a sound is different from the one you experience when you are looking at an object is that different parts of the brain are involved in receiving information coming in from the different sense organs. The neuronal circuitry ensures that the impulses from particular receptors end up at particular parts of the brain. In this way, the brain can discriminate between stimuli, even when they are close together. But each individual sense can vary too. Not only do our ears pick up sounds, but our brains can discriminate the pitch and loudness of those sounds from the information sent to the brain by the ears.

Perceptions, then, can vary both in quality and quantity. The quality of a stimulus might be, say, the flavour of a taste, the colour of light, or the pitch of a musical note. The quantity is its strength: whether the flavour is strong or the light is intense. As we shall see, sense organs, working with the brain, differentiate quality effects in many ways. But the strength of a stimulus is sensed along similar lines. A stronger stimulus makes bigger charge differences on the receptor cell membrane, which fires off a greater frequency of bioelectric impulses in the circuitry leading to the brain. A deafening sound is perceived as loud because the brain is bombarded with many impulses each second, far more than are generated by a gentle whisper.

Processing Information

Smell and Taste

In many ways, a sense of smell (technically, olfaction) is one of the most primitive of senses, one that appeared very early in the evolution of animals. Chemical odours detected by senses could signal nutritious food, hazardous poisons or (for smelly animals) dangerous predators or mates. Mammals, in particular, harnessed their sense of smell to the full as most of them evolved to be active at night, and smelling things was a good way of getting around. In day-active primates, especially humans, the sense of smell has diminished, but there are clues that point to its ancestral significance. We still have more than 300 genes concerned with smell distributed across 21 chromosomes, and it is one of the first senses to develop inside the womb. One study even suggests that we can detect upwards of a trillion different odour molecules. And the very first pair of cranial nerves in our head connects to our nose. Each of these olfactory nerves carries impulses to a part of the brain under the front of the cerebrum called the olfactory bulb. In most backboned animals, the olfactory bulbs are large, and often exposed in front of the brain, but in humans they are small and largely hidden. The olfactory bulbs process smells in much the same way that the thalamus at the core of the brain acts like the main hub for receiving other kinds of sensory information. They distinguish different odours and act as a kind of filter, enhancing some but screening out others. The bulbs then send impulses to a part of the cerebrum surface called the piriform cortex for processing. Interestingly, the impulses also go to parts of the brain involved in memory and emotion (the hippocampus and amygdala). This explains why smell, more than any other kind of sense, can trigger strong emotive memories. See chapter 9 for more about how smell links with emotion.

In contrast, the sense of taste seems to be much weaker than olfaction, even though three pairs of cranial nerves are responsible for carrying impulses from the taste buds on the tongue and the back of the mouth. Like most senses other than olfaction, taste impulses are sent to the cerebral cortex via the thalamus.

Receptor cells for detecting odour molecules are packed into the lining of the nasal cavity called the olfactory epithelium. Once stimulated, they send electrical impulses along sensory fibres into the olfactory bulb of the brain.

Olfactory bulb

Layer of receptor cells (olfactory epithelium)

Nasal cavity

Vision

Vision is the most important sense of all for humans, so it is no wonder that a large part of the cerebral cortex is devoted to processing the information that comes from our eyes. The back of each eye has a layer of light-sensitive receptors in the retina. When light shines onto these photoreceptor cells, impulses are generated that pass along the optic nerves – the second pair of cranial nerves – into the brain. Only the central patch of each retina is sufficiently sensitive to generate a high-resolution image, but the eyes scan the field of view with a series of jerky movements called saccades. This helps the brain quickly build up a complete detailed picture of the field of view. The eye's receptors in the centre of each retina come in three varieties, depending on whether they detect red, blue or green light. The combination of receptors stimulated helps the brain to perceive the final colour.

The optic nerves connect in a cross-shaped arrangement below the brain (beneath the hypothalamus) called the optic chiasm. This allows some of the sensory fibres to cross over and connect with the opposite side of the brain. The brain receives slightly different signals from each eye, meaning the cerebral hemispheres end up with a mixture of signals. From this mixture comes the brain's ability to perceive depth and three dimensions. Impulses pass via the thalamus to the visual part of the cerebral cortex, on the occipital lobe at the back of the head. It is here that the brain resolves details of the visual information collected by the eyes, such as shapes, sizes, colours and perspective. Some sensory impulses from the eyes pass first to a part of the midbrain called the tectum before then being processed by the thalamus. The tectum is involved in helping to control the eyes' saccade movements in building up a "map" of the scanned field of view.

A cross section through the head, looking down from above, shows that the left and right sides of the brain each receive electrical impulses from both eyes. This is because the sensory nerve fibres between the eyes and the brain meet at a cross-over point called the optic chiasm. It is this arrangement that helps us to perceive depth and three dimensions.

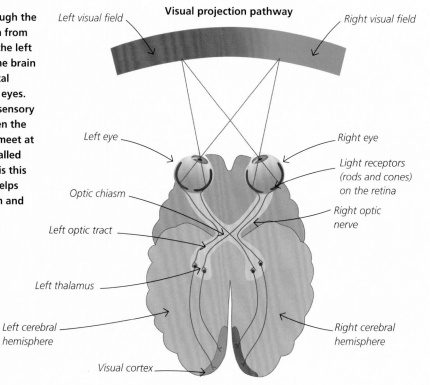

Visual projection pathway

Left visual field · Right visual field · Left eye · Right eye · Light receptors (rods and cones) on the retina · Optic chiasm · Right optic nerve · Left optic tract · Left thalamus · Left cerebral hemisphere · Right cerebral hemisphere · Visual cortex

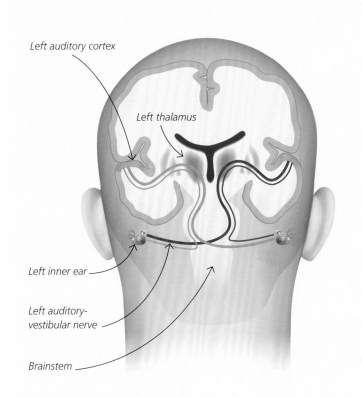

Left auditory cortex

Left thalamus

Left inner ear

Left auditory-vestibular nerve

Brainstem

A section through the head from the back shows the information pathways taken from the ears to the brain. Electrical impulses pass first through nuclei in the brainstem before passing up to the higher brain. Most impulses from the left side pass to the right side of the brain and *vice versa* (thick lines), but there are some impulses (thin lines) that stay on the same side of the head. This means that each half of the brain receives some information from both ears.

Hearing and Balance

Our ears perform a dual purpose. Some sensory cells in the inner ear detect sounds, while others are concerned with our sense of balance. The mechanism involved in each is similar: both involve sensors called hair cells.

Sound receptors are found in a part of the inner ear called the cochlea: a fluid-filled coiled tube that looks like a snail's shell. A ribbon of receptors runs through the tube right to its very centre. Sound waves funnelled into the ear first vibrate the ear drum. These vibrations are transmitted to the fluid in the cochlea by tiny bones. The fluid also vibrates, and this bends the hairs of the receptor cells. Vibrations produced by deeper-pitched sounds move further into the centre of the cochlea's coil. Sensory neurons are stimulated along the length of the cochlea, according to the pitch of the sound. This means the brain can tell the pitch of the

sound by detecting exactly where along the cochlea the impulses were generated.

The hair cells involved in balance are found in a neighbouring part of the inner ear called the vestibular apparatus. This consists of a system of semi-circular canals (three in each ear, arranged at right angles to one another) and chambers that are also filled with fluid and contain hair cells. Cells in the semi-circular canals detect how the head moves in three planes: front and back, side-to-side and twisting left to right. The cells in the chambers detect the position of the head. When combined, the information received by the brain helps to provide a sense of balance. Nerve fibres from both the cochlea and vestibular apparatus carry impulses to the brain stem through the eighth pair of cranial nerves, from where they are relayed to the thalamus and on to the cerebral cortex for processing.

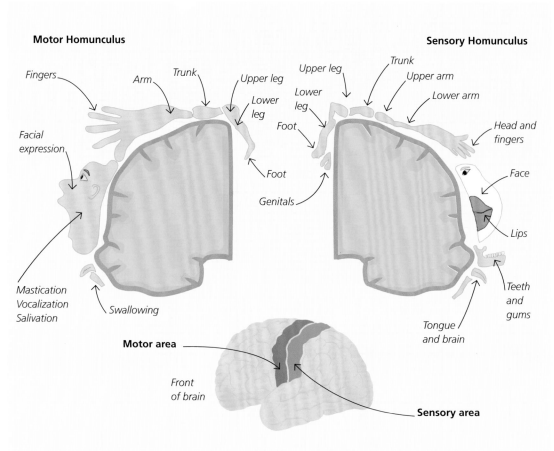

Motor Homunculus

Fingers

Arm

Trunk

Upper leg

Lower leg

Facial expression

Foot

Mastication
Vocalization
Salivation

Swallowing

Sensory Homunculus

Upper leg

Trunk

Upper arm

Lower leg

Lower arm

Foot

Head and fingers

Genitals

Face

Lips

Teeth and gums

Tongue and brain

Motor area

Front of brain

Sensory area

Maps or models of the human form, called cortical homunculi, have exaggerated body parts to reflect the relative roles played by the areas of cerebral cortex in processing information passing to and from these body parts. Sensory and motor impulses are controlled by neighbouring folds of the cerebral cortex.

Touch and Pressure

Receptors for sensing touch and pressure are found in the skin all around the body. Impulses from many of these receptors reach the brain via the spinal nerves and spinal cord; impulses from the head get there through several possible cranial nerves. Collectively, these make up the body's somatosensory system. After passing through the hub of the thalamus, this sensory information is passed on to the cerebral cortex.

This kind of sensory information is received and processed by a particular fold on the parietal (top-rear) lobes of the cerebral hemispheres: the

postcentral gyrus. The corresponding motor output comprises the signals sent to muscles so that they can respond. This is controlled by another fold just in front: the precentral gyrus on the front lobe. In the 1950s, American neurosurgeon Wilder Penfield mapped out the functional effects of these folds through his treatments of sufferers of epilepsy. By stimulating different areas of the gyrus and seeing which part of the body twitched, Penfield deduced how they were linked to different parts of the body. He presented the results in models of the human body with their parts proportionally enlarged according to the relative areas of the sensory and motor folds that control them. For instance, disproportionately large areas are devoted to sensations and movements of the fingers and hands, reflecting their sensitivity to touch and movement. These representations are called the cortical homunculi (homunculus being Latin for "little man").

Pain

Receptor cells that detect pain are called nociceptors. Pain is regarded as any stimulus that is sufficiently intense to cause potential harm. Impulses produced by pain move more quickly than other impulses so they reach the thalamus within a very short time. From there, they pass on to many different parts of the cortex. Unlike other senses, pain is not represented in a specific location..

The sensation of pain is the body's punishment mechanism. It evolved to prevent the body from repeating any behaviour that is likely to cause harm.

Sixth Sense

Somewhere between our ability to sense our surroundings and our sense of what is going on inside our bodies is a strange sixth sense called proprioception. This is the sense that tells us the positions of our body parts and what they are doing. It is easy to take proprioception for granted, but it is critical for keeping our sense of position and balance. For instance, even if we are blindfolded, we know the positions of our limbs in relation to our body. About 70 per cent of the information needed comes from sensors all around the body. They send impulses to the brain, indicating how and where the muscles are contracted, tendons are stretched, skin is compressed, and so on. At the same time, the inner ears and eyes send impulses about the position and movement of the head, helping with the sense of balance. The thalamus, the sensory hub at the centre of the forebrain, is involved in collecting this information. But our automatic, unconscious control of proprioception is largely performed by the cerebellum, the part of the brain concerned with automatic coordination of complex movements.

Signals from receptors in the arms and legs tell your brain exactly where they are, no matter how awkward their position.

Information Flow through the Brain

The brain is more than just a relay centre for information. It also stores information as memories and coordinates impulses to create new information.

There is no single part of the brain that is solely devoted to receiving all the sensory information. Impulses from sense organs travel up through the spinal cord or along the sensory fibres of the cranial nerves, and they can affect lots of different parts of the brain. Once they reach the brain, there are many centres – nuclei of neurons – that are specialized in handling the input of information, These centres then have the job of generating more impulses, which are sent through association regions for coordinating this information. Finally, impulses are sent from the association regions along motor fibres to muscles and glands, either back out through the spinal cord or taking a different route along cranial nerves.

Within the association regions of the brain, information received is processed to generate an appropriate response. This involves decision-making processes that ensure that impulses are transmitted along one route instead of another. These decisions can be affected by memories stored in the brain, making the association regions of the brain some of the most complex, both in terms of the arrangement of neurons and what is happening in them.

Nerve cells stretch throughout the body to create an intricate network that carries signals with astonishing speed.

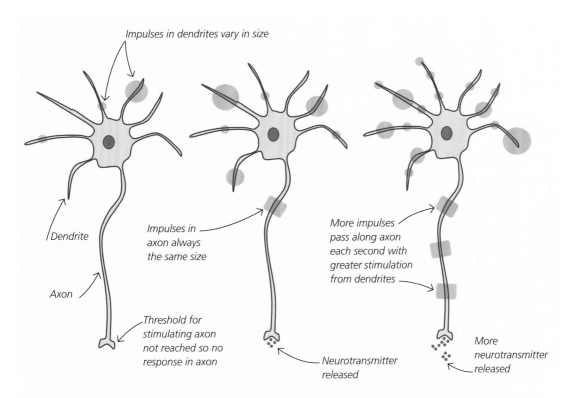

Impulses in dendrites vary in size

Dendrite

Axon

Impulses in axon always the same size

Threshold for stimulating axon not reached so no response in axon

Neurotransmitter released

More impulses pass along axon each second with greater stimulation from dendrites

More neurotransmitter released

A typical neuron in the brain effectively funnels electrical impulses of dendrites collected from many different synapses into a single long fibre called an axon. If the combined effect of these impulses is strong enough, an impulse of a fixed size called an action potential is formed in the axon.

How Brain Cells Carry Impulses

The brain cells involved in carrying impulses are neurons. And the vast majority of neurons in the brain are association neurons concerned with coordination and control. These have lots of spindly fibres for communicating with the multiple fibres of neighbouring cells. In fact, a single neuronal brain cell could have thousands of such fibres, opening up thousands of possible circuits for the impulse to take. As we have seen (chapter 2), impulses move through neurons in one direction. Typically, one fibre, called an axon, has the job of carrying the impulse away from the cell body, while others, called dendrites, funnel impulses towards it. These multiple incoming impulses could have been generated at several synapses at once, and they converge towards the cell body before merging into the axon.

Some of the impulses generated in the dendrites are stronger than others: their strength depends on the amount of stimulation that produced them. When these impulses arrive at the neuron's cell body, all their effects are combined to stimulate the axon leading away to the next synapse. In order for the signal to continue down the axon, the combined effect of all these impulses must be big enough to excite this axon. Each axon that gets sufficiently excited can then transmit this impulse right down to the synapse at its tip. Here the impulse triggers the release of a neurotransmitter, which can stimulate a new impulse in the dendrite of a neighbouring neuron, and so on. Signals pass through each neuron of the brain in this way, from dendrite to axon and synapse.

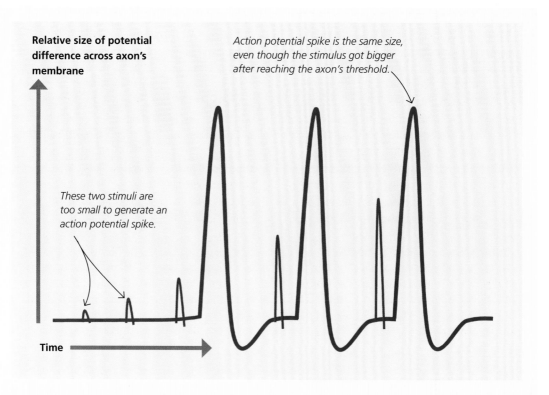

Relative size of potential difference across axon's membrane

Action potential spike is the same size, even though the stimulus got bigger after reaching the axon's threshold.

These two stimuli are too small to generate an action potential spike.

Time

The electrical impulse carried on a neuron's axon is recorded as an action potential spike (red) on a trace of the neuron's activity. The spike is always the same size, but the stimulus (blue) must be above a certain threshold to generate it.

Neuron Spikes

When the level of stimulation picked up by a neuron reaches a certain necessary threshold, it sends an impulse down its axon. The size of the electrical signal in the axon's impulse stays the same, no matter how big the stimulation, as long as it reached the minimum threshold. If the stimulation is too weak and below the threshold, the axon does not send an impulse at all. This helps to prevent the nervous system from getting overstimulated by the slightest provocation. Axons always carry impulses of about the same size: typically about 50 millivolts. This is tiny, but more than enough for it to trigger the bioelectric impulse (see chapter 2). The axon impulse is technically called an "action potential". But it is also popularly called a "spike", as it produces a noticeable

spike-like peak in the traces produced by electronic equipment that measures the electrical activity of the brain.

As we have seen, the brain discriminates between strong and weak stimulation by the frequency of impulses (spikes) generated. Thousands of spikes shoot through the axons of brain, with areas of rapid-firing representing the regions of more intense stimulation.

Combining Information

At any one time, whether awake or asleep, your brain is busy ferrying millions of spikes at once. But for brain cells to perform all the complex tasks that are necessary in a clever, working brain, they need to do a lot more than just pass information from one neuron to another. They must be able to consider many different sources of information, including the information input coming from senses and information stored as memories. In other words, brain cells must integrate all this information before generating an output that can cause a response.

Integration of information in the brain is made possible by the ways neurons carry signals, and by the different effects of neurotransmitters. Neurotransmitters are chemical signals that have different effects on target neurons, largely depending upon the receptors that intercept them. Some brain neurotransmitters, such as glutamate, are exciters: they stimulate new impulses in adjacent neurons; others, such as GABA, are inhibitory, and stop neurons from getting electrically excited (see chapter 6). Neurotransmitters from lots of cells can converge into the dendrites of one neuron, where the overall effects are combined. For instance, an abundance

of glutamate from many synapses would produce a strong impulse response in the dendrites of a neuron, and so be big enough to successfully fire a succession of spikes in its axon. But if a neuron were stimulated by a combination of glutamate and inhibitory GABA, fewer spikes would be generated in the axon, or even none at all. This is because the inhibitory effects of GABA subtract away from the excitatory effects of glutamate. And this, in turn, would determine the amount of neurotransmitter released from this neuron, affecting neighbouring neurons, and so on.

Integration of information in brain circuits in this way therefore depends on the balance between neurotransmitters that excite and those that inhibit, as well as the pattern of synapses that physically allow communication between one neuron and the next. Over time, certain synapses that are over-used become strengthened (a process behind memory storage), which also influences the flow of signals.

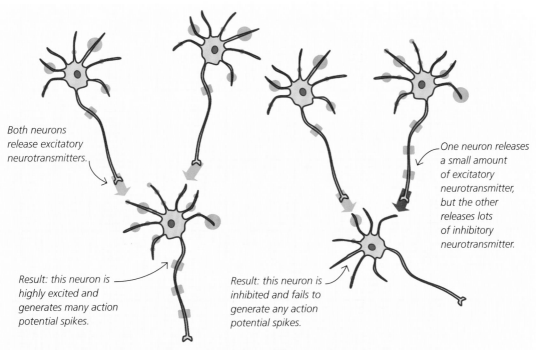

Both neurons release excitatory neurotransmitters.

One neuron releases a small amount of excitatory neurotransmitter, but the other releases lots of inhibitory neurotransmitter.

Result: this neuron is highly excited and generates many action potential spikes.

Result: this neuron is inhibited and fails to generate any action potential spikes.

The behaviour of the brain's neuron's at any one time and place depends upon the combined effects of chemical neurotransmitters that excite or inhibit.

Spontaneous Brain Waves

Remarkably, neurons may continue to fire spikes and release neurotransmitters even without any input of sensory information. Spontaneous activity of electrically excitable cells is, in fact, nothing new. The heart, for instance, is made of specialized muscle that spontaneously twitches in a rhythmic fashion, without any stimulation from the nervous system. (Remember that the nerves that connect to the heart are involved in regulating its rate, not triggering the heartbeat itself.) Some nuclei in the brain contain pacemaker neurons that have rhythmic activity, giving rise to so-called brain waves. Some may play a role in controlling rhythmic movements, such as breathing or walking. Others are involved in higher functions and help with integrating information.

Brain Pathways

Brain neurons with short axons carry their spikes to their closest neighbours, and these are the ones most involved with information-processing. However, neurons with the longest axons carry spikes between entirely separate parts of the brain. These are responsible for the brain's internal communication network that carry spikes to and from the processing sites.

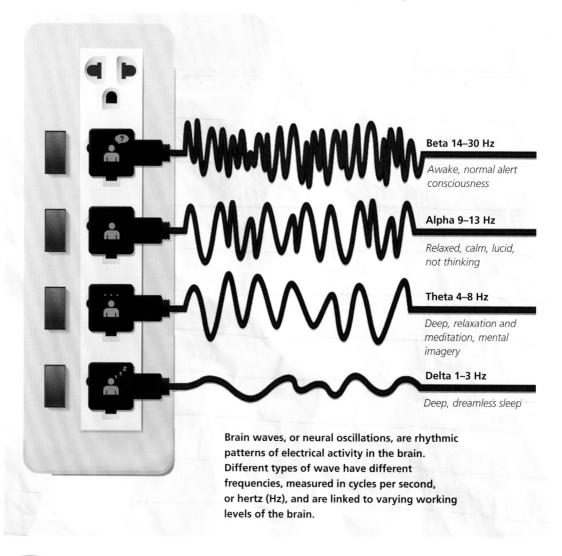

Beta 14–30 Hz

Awake, normal alert consciousness

Alpha 9–13 Hz

Relaxed, calm, lucid, not thinking

Theta 4–8 Hz

Deep, relaxation and meditation, mental imagery

Delta 1–3 Hz

Deep, dreamless sleep

Brain waves, or neural oscillations, are rhythmic patterns of electrical activity in the brain. Different types of wave have different frequencies, measured in cycles per second, or hertz (Hz), and are linked to varying working levels of the brain.

Tracking Brain Circuitry

The electrical activity of the human brain can be measured by placing electrodes on the scalp – a technique called electroencephalography (see chapter 2). But this technique can only monitor the effects of millions of neurons at a time. This is good for seeing how parts of the brain are activated during certain behavioural tasks, but no good for studying how individual neurons interact and communicate.

For more detailed studies, more invasive methods are needed, and our understanding of fine-level brain function has depended a lot on experiments with animals. Devices called multi-electrode arrays, made up of clusters of up to hundreds of electrodes, are inserted into the brain so that each electrode can monitor the activity of individual neurons. This can produce a trace that shows the activities of many neighbouring neurons over long periods of time, helping scientists to work out the circuitry patterns that exist in parts of the brain. Another technique involves injecting the brain with fluorescent dyes that stick to particular kinds of brain cells, and which show up electrical activity as flashing spots. But by far the most precise technique, called patch-clamping, involves placing electrodes inside individual neurons to detect cell behaviour, even when at rest.

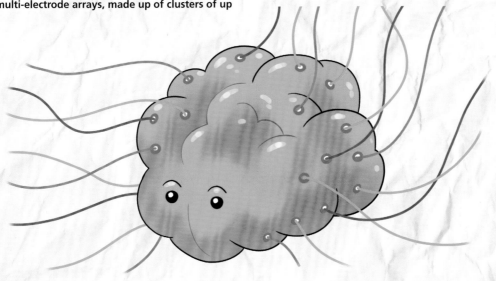

In the deepest parts of the brain, the information-processing sites are centred on regions called nuclei, which, as we have seen, are concentrations of grey matter consisting of many cell bodies. Many of these nuclei are clustered together in parts of the brain such as the thalamus, hypothalamus and medulla (see chapter 5). But in the higher parts of the brain, including the cerebrum, as well as the cerebellum, the grey matter forms a continuous surface layer: the cortex. Such a large expanse of cell bodies (as well as their short axons) reflects the higher demands of the kinds of higher-level thinking involved here. Communication between nuclei, and between nuclei and the cortex of the cerebrum and cerebellum, happens along bundles of longer axons called tracts, which make up the white matter of the brain. Tracts help the control centres in the core of the brain, such as the thalamus, interact with higher parts of the brain involved in conscious thought: the cerebral cortex. Others run through the corpus callosum, the band of tissue that connects the brain's hemisphere's, allowing right and left sides to communicate.

The Cerebral Hemispheres

The two halves of the main dome of the brain, the cerebral hemispheres, make up the biggest part of the human brain. They are critical for the higher functions of the brain.

The cerebral hemispheres are the most instantly recognizable parts of the human brain. Together they make up the higher part of the brain called the cerebrum. They all but envelop the brainstem underneath and the bulging cerebellum at the back. The two hemispheres have a wrinkled surface and are virtually symmetrical, except that the right side is slightly bigger and is very slightly warped further forward than the left. Each hemisphere contains a ventricle filled with cerebrospinal fluid, and deep nuclei of grey matter overlain with white matter. Finally, there is surface grey matter (the folded layer) on top. A region of white matter called the corpus callosum connects the two hemispheres to one another.

Contrary to popular thought, most of the higher-level functions of the cerebrum, such as sensory perception and memory, are divided between both hemispheres, but there are certain specific differences in how they work. Notably, centres for producing and understanding language are confined to the left side, but there are some

Right and Left Brains

In 1981, American neuroscientist Roger Sperry (1913–1994, right) won a Nobel Prize for his research into the idea of the split brain. His work involved tests on patients who had had the fibrous connection between their hemispheres, the corpus callosum, severed as part of a surgical treatment for epilepsy. In his experiments, Sperry presented patients with words or objects that could only be seen on the right or left sides of their body. Because the visual field on one side is controlled by the hemisphere on the opposite side, Sperry was able to deduce how the two hemispheres differed in their processing skills. He discovered that the left hemisphere was more concerned with verbal skills and analysis, while the right hemisphere was involved with context, tone and comparison. Sperry's work led to the notion that the left side was the logical, scientific brain, while the right side was imaginative and artistic. Today we know that skills in all these disciplines involve both sides, but the hemispheres do have some functional differences in the way Sperry concluded. A study by the University of

Utah in 2013 showed that, there are similar amounts of activity on both sides of the brain regardless of a person's skills or personality.

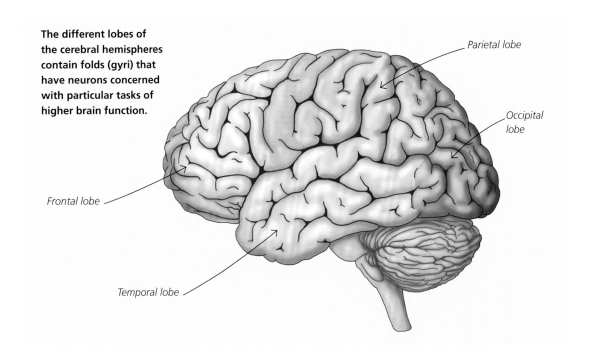

The different lobes of the cerebral hemispheres contain folds (gyri) that have neurons concerned with particular tasks of higher brain function.

Parietal lobe

Occipital lobe

Frontal lobe

Temporal lobe

general chemical differences too. For instance, pleasure-inducing dopamine is produced in greater quantities in the left.

Organization of the cerebrum

The cerebrum starts to develop in the embryo at around five weeks, and forms as the sides of the rudimentary brain inflate outwards. Each hemisphere envelops a fluid-filled ventricle at its centre. On each side, the wall of the hemisphere folds inwards at the top and bottom and grows into the developing brain. These folds end up within the brain and both become deep regions of grey matter: the lower ones develop into so-called "basal" nuclei of the cerebrum, while the upper ones form a structure involved in memory storage called the hippocampus. Meanwhile, the outer walls become the surface layer of grey matter.

In primates, more than 70 per cent of neurons of the entire central nervous system end up developing in the outer wall of each cerebral hemisphere. Specifically, a patch of wall near the top of the embryonic cerebrum expands outwards to encompass virtually the entire surface of each

hemisphere. As it does so, the walls become folded to form the grooves and wrinkles that are characteristic of the mature cerebral hemispheres: this is the cerebral cortex (also called neocortex). The folding happens so that a greater area of cerebral cortex can be accommodated within the confines of the skull. The cerebral cortex is the main part of the brain involved in deciphering sensory information. While other parts of brain, such as the thalamus, act as hubs for receiving information, the cerebral cortex carries out many of the functions involved in coordinating it.

The fixed way the cerebrum grows during normal development means that its shape always consists of a number of distinct lobes. On the surface, five distinct lobes develop, some named after the skull bones under which they are located: frontal, occipital (back), parietal (top-rear) and two temporal lobes (on each side). Each lobe contains neurons that are involved in controlling specific aspects of sensory perception or motor activity. A small region beneath the cerebral hemispheres, called the piriform cortex, is specifically involved in sensing smell.

Regions of the Cerebral Cortex

Towards the end of the nineteenth century, German scientists invented techniques that enabled them to reveal different kinds of neurons under the microscopic by staining them. The German neurologist Korbinian Brodmann used these staining methods to produce "maps" of the neurons in the cerebral cortex of different animals. In 1909, Brodmann published the first such map of the human cerebral cortex. In it, he identified 47 brain regions that differed according to the types of neurons found there and how they were layered. These regions are now known as Brodmann areas, and they are still used, in a refined form, as the basis for mapping the cerebral cortex (50 are now recognized for a human brain.) Studies since Brodmann's time have shown that many of his areas are linked to specific functions of the brain.

Distinct areas of the cerebral cortex have sensory, association or motor functions. The cerebral cortex of the parietal, temporal and occipital lobes are involved in receiving and processing sensory information. For instance, the cortex of the occipital lobe is concerned with vision. The frontal lobe is mainly involved in controlling motor activity: in other words, conscious movements that relies on contracting voluntary muscles. The front part of the lobe, the prefrontal cortex, is also involved in so-called "executive" functions, which help the brain to differentiate and decide between potentially conflicting thoughts: here, many aspects of social behaviour, planning and personality are determined.

Central Parts of the Cerebrum

As we have seen above, where the upper and lower folds of the cerebrum grow inwards in the

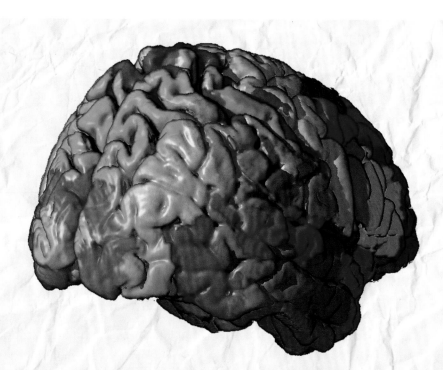

Since the pioneering work of Korbinian Brodmann in the early twentieth century, most of the regions of the cerebral cortex that he mapped by their structure and arrangement of neurons (here shown by different colours) have been shown to have specific functions.

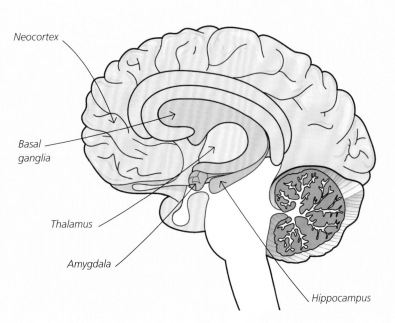

Deep inside each cerebral hemisphere, positioned just outside the hemisphere's thalamus, is an arrangement of structures involved in a number of functions, including memory and emotion. It makes up part of the brain's limbic system.

Neocortex

Basal ganglia

Thalamus

Amygdala

Hippocampus

embryonic brain, these parts become centralized areas of grey matter that are separate from the cerebral cortex.

Each of the upper folds goes deep into the cerebrum and becomes a part of the brain called the hippocampus: each hippocampus comes to lie on the inner side of a temporal lobe. "Hippocampus" means "seahorse", a name given to it for its vague resemblance to the shape of this animal. Like the rest of the cerebrum, the hippocampus receives sensory information, but it is mainly concerned with storing memory (see chapter 8). It is also involved in generating inquisitive or investigative behaviours and plays a central role in navigation.

The lower folds of the cerebrum become a cluster of nuclei called basal ganglia. These receive sensory information from the tegmentum part of the brain's core (midbrain: see chapter 5) and they help to control movement by monitoring the body's position and motivational state. In this way, they can select movements that are appropriate and inhibit others that are not. When the information flow between the basal ganglia and tegmentum is disrupted, the result is involuntary tremors. This is

what happens in Parkinson's disease. Other nuclei in the lower parts of the cerebrum are clustered together on each side to form a structure called the amygdala, which among other things is concerned with controlling emotion (see chapter 9).

Language

When Korbinian Brodmann mapped out the surface regions of the cerebral hemispheres, he did so based on anatomy alone: the Brodmann areas were identified by painstaking examination of brain tissue under the microscope. Brodmann suspected that his map might one day be a spatial plan of the brain's functions, but he knew that much more research would be needed to achieve that goal. In fact, neuroscientists had already learned a great deal about how different parts of the cerebral cortex were linked to function by looking at the symptoms of people suffering from brain injury and disease. Just as early anatomists had learned about functions by deliberately disabling parts of an animal's nervous system and looking at the effects, neuroscientists studied brain injuries as though they were natural experiments.

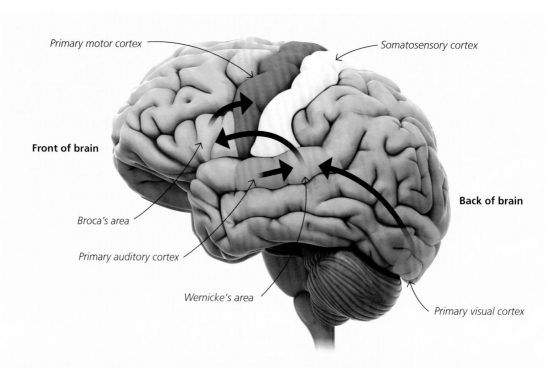

Primary motor cortex

Somatosensory cortex

Front of brain

Back of brain

Broca's area

Primary auditory cortex

Wernicke's area

Primary visual cortex

Wernicke's area in the cerebral cortex receives information from visual and auditory regions of the brain and enables language to be understood and interpreted. It then passes impulses to Broca's area, which converts this information into motor commands, allowing the person to speak.

In 1861, a French physician called Paul Broca heard of a patient at a Parisian hospital who had a severe language disorder. Louis Victor Leborgne was nicknamed "Tan" because that was the only word he could speak clearly. Remarkably, however, his understanding of speech and overall mental ability were not impaired. When Leborgne died, Broca performed an autopsy. Leborgne had been suffering from a lesion in the frontal lobe of the left cerebral hemisphere. Broca went on to discover a similar disorder in other patients, and the region he identified became known as Broca's area. It corresponded with two areas (numbers 44 and 45) that would be described by Korbinian Brodmann nearly half a century later.

A contemporary of Broca, German physician Carl Wernicke, had a similar interest in the link between brain injury and language disorder. Wernicke recognized that language disorders could also arise from damage to another part of the brain

near the top of the temporal lobe (later described by Brodmann as area 22). For people with damage to this part of the brain, their language sounded fluent and natural, but it lacked meaning. They had the ability to articulate the words, but not to understand them. Wernicke's area was therefore concerned with language comprehension rather than production. More recent research points to Wernicke's area being a more general sound-comprehension region, since patients with damage to this area may also have difficulty understanding non-verbal sound such as animal noises.

Both Broca's and Wernicke's areas have been linked to the left side of the brain. This is regarded as the dominant hemisphere in most people, but there are exceptions. In 5 per cent of right-handed people, the dominant hemisphere is on the right, so this is also the location of their language-processing areas. In left-handed people, right-hemisphere dominance rises to 30 per cent.

Processing Information

Chapter 8
MEMORY

The Power of Memory

Memory allows us to learn from past experiences, meaning that we can modify what we think and do. This frees us from the restrictions that were placed on our brains by our genes.

Some of our behaviour is so deeply programmed into our being that we have little choice in the matter. We shiver when we are cold, sneeze when something irritates our nose, and if somebody taps our knee in just the right place our leg shoots forwards. The programmes concerned with these actions are not only hard-wired into the circuits of our nervous systems, they are programmed into our genes long before the nervous system has even matured. As we grow from fertilized eggs and embryos, the genes in our cells determine the way the nervous system develops, so that those circuitries are laid down in certain ways. Behaviour that is genetically set in place is described as innate: it can be executed without any prior experience or learning.

The simplest kinds of animals are dominated by this kind of innate behaviour. They respond automatically to stimuli that are predetermined during development. Unless there is a problem with its internal wiring, for instance, a dragonfly nymph will always crawl out of a pond before it changes into a dragonfly. Innate behaviour can persist in more complex animals, especially when they are very young. A bird chick automatically begs for food, just as a human baby will automatically suck on a teat. But as bigger brains mature, they can do more to influence behaviour: they can remember and learn. And as bigger brains accumulate more memories, they can learn more things to help the conscious parts of their brains go far beyond their primitive innate behaviours.

Learning and remembering how to make a simple tool to extract termites from hard-to-reach places can give a bonobo an advantage over other animals in its environment.

They can add new learned behaviours to the behavioural repertoire. And learning helps us experience emotion and intelligence.

Why is Memory Significant?

Any animal that can remember things well has an advantage when it comes to their fight for survival. Remembering good things, such as a source of tasty food or where to find a mate, promises the possibility of nourishment or passing on genes. And remembering bad things, such as a near-miss with a dangerous predator, could save your life. Memory makes behaviour versatile and helps it to adapt to new circumstances. Without memory, behaviour would be quite routine and predictable. Memories are stored as physical connections between neurons in the brain (see next section), and bigger brains, with more neurons, can store more memories. As long as these connections are preserved, a longer life can also accumulate more memories over a longer period of time. Unsurprisingly, therefore, the animals with the best memories have the biggest brains and tend to live the longest.

Genes: the Root of Behaviour

Ultimately, the roots of our behaviour can be traced back to the genes inside our cells. These packets of information, made from DNA, determine the way cells grow and develop. As a single fertilized egg cell grows into an embryo made up of billions of cells, genes in different parts of the body are switched on or off, depending upon where they are. As a result, the body develops distinctly different parts with separate organs, and so on. Brain cells contain the same genes as any other cell around the body, but only have the genes for brain functions activated.

Specifically, genes control the proteins that are made by cells. Proteins encompass a very wide range of different kinds of molecules that perform many different tasks. Brain proteins, for instance, are responsible for generating the electrical charges on neuron cell membranes that allow them to carry impulses. They are also receptors that intercept neurotransmitters: the chemical signals

The human brain, with its enormous cerebral hemispheres, sitting inside a body that can live for decades, has a particularly impressive ability to store and recall information.

that help adjacent neurons communicate with one another. Other receptors bind to hormones circulating in the blood stream. Other proteins make up the scaffold for building neurons (and their fibres) themselves. This rich mixture of different types of proteins, encoded by inherited genes, is ultimately responsible for ensuring that a brain performs in the way it does at different times. And different brains from different animals, or even different people, have their own idiosyncrasies.

Although specific memories and learned behaviours themselves are not genetically determined, the ability to remember *is*, like any other basic aspects of brain function. Genes that encode for bigger, more complex brains will also provide the ability to store more memories and learn from experiences.

When a baby is touched on the cheek, it will automatically turn its head towards the touch with an open mouth. This rooting reflex is not learned but innate, and helps the baby find the teat for getting milk.

Innate Behaviour

The very simplest type of behaviour is innate. This is any pre-set behaviour that doesn't require any degree of learning or practice at all. Innate behaviour is usually stereotyped: this means that it happens in much the same way every time it occurs. Innate behaviour predominates in animals with simple brains that have little capacity for storing memories. When an earthworm digs down into the safety of a dark burrow or a butterfly visits a colourful flower to drink nectar, this is innate behaviour. The behaviour is essentially "hard-wired" by the animals' nervous systems, in turn fixed by the inherited genes. This means that the same kind of innate behaviour is likely to be shared by all members of a species because they share similar sets of genes. And it is passed down through generations with the genes. It evolves by natural selection, whereby certain sets of genes improve survival and are more likely to be passed on by reproduction. This is responsible for the huge variety of innate behaviours found in animals alive today.

But innate behaviour does not solely belong to animals with the very simplest brains. It persists in big-brained animals too, including ourselves. The most straightforward kinds of human innate behaviour rely on reflex actions that involve impulses passing through parts of the central nervous system that happen automatically, without the interception of the conscious parts of the brain. The knee-jerk reflex is an example. Other movements involving more complex pathways and higher-order processing, but that are still automatic reflexes, are the primitive reflexes of babies. For instance, the sucking reflex, which is triggered by touching the roof of the mouth, helps the baby drink milk, while the rooting reflex, when the baby turns to face something that strokes its cheek, brings the baby's head into position for sucking.

Effects of the Environment

In reality, of course, the final characteristics of any adult living thing are not just determined by inherited genes. As the body develops, it is exposed to different factors from its surroundings, which affect the progress of development. More food, for instance, can clearly affect how the body grows. Characteristics related to the nervous system and behaviour are no exception, and memory plays a central part in this.

As the developing body is exposed to different kinds of sensory stimuli, brain cells build up specific connections that are preserved as memories. These memories then affect behaviour. This is the essence of learning. While the capacity for learned behaviour is genetic (in the sense that genes determine the complexity of the brain), the actual behaviour that is learned is not. This means that even individuals that are genetically identical can end up with very different learned behaviours. In contrast to innate behaviour, therefore, specific learned behaviour is not genetically inherited, not stereotypical, and cannot evolve by natural selection. The only way learned behaviour can be passed on is through parents teaching their offspring, or by cultures passing on their traditions.

In reality, most types of behaviour are a complex mixture of innate and learned aspects. Some reflexes, for instance, can be modified through experience. And even the simplest animals have some ability to store memories in their nervous systems, although these are limited by the size of the brains. But for a complex human brain, the relative contributions of these innate and learned aspects can often be difficult to disentangle. The relative influences of genes and environment on human behaviour, such as various expressions of intelligence (see chapter 10), form the basis of an ongoing nature–nurture debate.

Types of Learning

Learned behaviour covers any kind of change in the behaviour of an individual as a result of experience. The change is usually adaptive:

"adaptive change" is something that happens to the advantage of the learner. For instance, by remembering a painful wasp sting, you will learn to avoid black and yellow insects in future encounters: that's a good thing if you want to avoid a painful sting. In humans, learning relies heavily on "higher" parts of the brain, such as the cerebral hemispheres, where memories are stored. Effective learning, at least in animals, often depends upon on whether a stimulus is beneficial, harmful or neutral. Beneficial or harmful stimuli offer reinforcement: it is clearly adaptively valuable to learn to take advantage of a beneficial stimulus (a reward), but to avoid a harmful one (a punishment). The rewards and punishments are reinforcements. Likewise, it is adaptively valuable to learn to ignore a neutral stimulus, because responding to it would be a waste of time and energy. For this reason, the memory systems of the brain are intimately linked with parts involved in controlling sensation and emotion (see chapter 9).

Even though identical twins are genetically identical, they can each learn to perfect very different skills. This is because their brains are not entirely hard-wired. As the twins grow, their brains can develop different connections which enables them to learn different things.

Through classical conditioning, we learn to avoid any insect with yellow and black markings following a nasty encounter with a wasp.

One of the simplest forms of learning is called habituation. This happens when we learn to snuff out a response when a stimulus is repetitive and neutral: in other words, it serves no useful purpose. It's an advantage to learn to ignore the stimulus because it saves energy, lowers stress and means more time can be spent on other more useful activities. Habituation can lead to the body becoming permanently unresponsive to a stimulus or combination of stimuli. Birds learn to ignore a scarecrow by habituation. In the same way, our brain learns to ignore the sensation of clothes against our skin that would otherwise be continually firing impulses from touch receptors. Habituation depends on the fact that neurons can become desensitized when they are overly stimulated: with persistent stimuli, they no longer release neurotransmitters, so signalling stops. Some receptor systems, such as the sense of touch, habituate more easily than others, such as those involved in pain. That is why we can quickly learn to ignore a shirt on the back, but not an abscess in a tooth.

A more complex learning behaviour is involved when a neutral factor is associated with a reward or punishment. This ends up either encouraging a repeat performance or stopping the behaviour from happening again. This is called conditioning. The neutral factor could be stimulus, in which

case it is called classical conditioning. It happens with our encounters with the stinging wasp. The bright yellow and black markings of a wasp are themselves harmlessly neutral, but we learn to associate them with the punishment of a sting. We become conditioned to avoid any buzzing insects with these warning colours.

Sometimes the neutral factor is something that is done by chance. This kind of learning is called operant conditioning. It starts with some kind of behavioural operation, and typically involves trial and error. If a bird disturbs some leaf litter by chance and finds a tasty worm, it will learn to purposefully turn over leaves again. The power of operant conditioning is well known in parental management of children, where good behaviour is rewarded and bad behaviour is punished.

The most advanced learning does not involve conditioning or trial and error, but relies on higher mental processes such as thought and reasoning. This is called insight learning, and is used to solve problems based on numerous incoming sources of information. For this to work, the brain must be able to integrate the information effectively to make the best decision. Insight learning has been recorded in birds and mammals, such as the use and modification of tools, but is notably well advanced in apes and especially humans.

How the Brain Remembers

The higher parts of the brain, the cerebral hemispheres, have areas that are devoted to storing memories. When this happens, the activities of neurons affect the ways they connect at synapses, so experiences end up leaving a physical impression on the brain.

The California sea hare, *Aplysia californica*, from the Pacific coast of North America is a species of giant sea slug that grows to the size of a kitten. It looks about as unlikely to reveal the secrets of the human brain as it is possible to imagine, but this strange animal has given scientists an extraordinary insight into the mysteries of memory and learning. In the mid-1960s, the Austrian-born neuroscientist Eric Kandel chose the sea hare to study memory and learning at the level of individual neurons, something that had never been attempted so thoroughly before. The sea hare was ideal for this task. Its central nervous system contains only about 20,000 neurons (a tiny fraction of the billions in human brains), and the individual neurons are comparatively huge, meaning that they could be more easily observed and manipulated. Kandel and his team not only demonstrated how a sea hare's neurons changed when they stored memories, but also how they were used to help the slug to learn simple things, such as avoiding an irritating poke with a stick.

Kandel's ground-breaking work on the physiological basis of memory earned him a Nobel Prize in 2000, and serves as the basis for the current understanding of how memories are stored in our brains.

The Physical Basis of Memory

The work of Kandel and his associates revealed that memory-storing neurons produce particular kinds of proteins, which affect the way the neurons work. Specifically, they studied the response of the sea hare's neurons to repeated stimulation of its gills, a particularly sensitive part of the animal's body. Over time, the slug was clearly remembering the stimulation and learning to react to it with a bigger response. Kandel identified a chemical effect, whereby the release of the neurotransmitter serotonin set up a chain reaction that ultimately affected the charges on the membranes of synapses. Repeated stimulation of synapses in this way was activating proteins that helped to reinforce existing synapses and encourage the formation of new ones.

The California sea hare is a popular animal in studies of the brain and memory. It has unusually large brain cells with cell bodies up to 1 mm across, and their arrangement is far simpler than in a human brain. This allows neuroscientists to observe the effects of memory and learning directly on the brain's neuron circuitry.

Because of the simplicity of the sea hare's central nervous system, Kandel was able to see the effect of these memory-learning experiments on the way individual neurons are interconnected at their synapses. Basic neuron circuits are set in place during development of the embryo, as dictated by the cell's genes. But over time, the sea hare's mature circuitry could be altered as new synapses replaced old ones and some existing ones were reinforced. In other words, the arrangement of neurons is said to be plastic. This synaptic plasticity is the physical basis for storing memories. Over time, memories are stored as a "map" of interconnected neurons that are imprinting new neuronal circuits on the nervous system.

Short-term Memory

Before memories can be stored in the long term, events must influence neurons in the short term. Neurons change very quickly – within fractions of a second – when they conduct impulses, and take only slightly longer when they release neurotransmitters. The pattern of electrical activity and distribution of neurotransmitters can be looked upon as a trace of the stimulus event. But as neuron membranes recover and neurotransmitters are depleted (see chapter 2), the trace is only fleeting. The fleeting effects on neurons make up short-term memory. They last no more than a minute before they are gone. Short-term memory is involved in the moment-to-moment conscious processing of the sensory information: it is "working" memory. But its capacity is small. Current research suggests that no more than four items can be accommodated by the short-term memory at any one time, although, at least for humans, this is complicated by the effects of something called "chunking".

Chunking happens when individual items in the short-term memory are combined into meaningful chunks according to how a person thinks. For instance, the four words "Thursday", "on", "wasp" and "stung" could be combined into the single meaningful chunk "wasp stung on Thursday". This effectively now becomes a single incident-related item for memory storage. The ability of humans to use their linguistic skills to chunk in this way can make it difficult to objectively compare short-term memory storage in different cultures or in different kinds of animals. But no matter how short-term memories are stored, their faintness means many are forgotten:

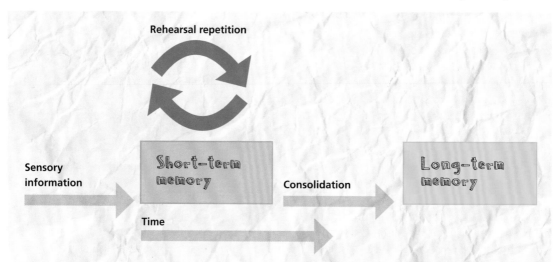

Rehearsal repetition

Sensory information

Short-term memory

Consolidation

Long-term memory

Time

Many of the short-term memories that we build up in our brain are lost, but over time some are retained as long-term memories. Short-term memories that consolidated in this way do so when they are repeated.

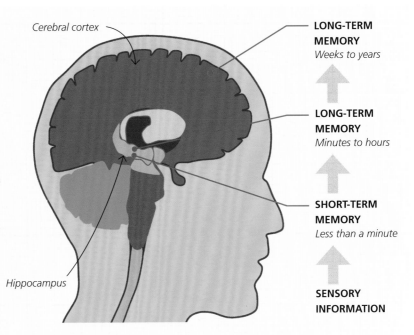

Memories are initially processed in the hippocampus, a part of the brain that lies deep within the temporal lobe of each cerebral hemisphere. This is where new patterns of synaptic connections are being assembled as short-term memories are changed into long-term ones. Over a much longer time scale, these memories are eventually imprinted in the cerebral cortex and stored there independently of the hippocampus.

Cerebral cortex

Hippocampus

LONG-TERM MEMORY
Weeks to years

LONG-TERM MEMORY
Minutes to hours

SHORT-TERM MEMORY
Less than a minute

SENSORY INFORMATION

they will fade just as their bioelectric and chemical effects disappear when neurons automatically recover from their stimulation.

Long-term Memory

The fleeting nature of short-term memory would render it useless unless its effects could somehow get translated into something more permanent, and this is exactly what happens. Kandel's work on the California sea hare showed how it is possible for short-term memories to turn into long-term ones. When a stimulus has a continued impact, such as through repetition, some of the subtle bioelectric and neurotransmitter effects are reinforced. This strengthens the synapses involved, helping to preserve the short-term memory traces in the long-term physical circuitry of the brain. This is called memory consolidation. The effects last longer – potentially much longer – than a minute, and memories become long term. As we know, the strongest human memories can last a lifetime.

Storing Memories

Theoretically, any part of the brain might be capable of storing memories. Wherever neurons are plastic, in other words wherever their synaptic connections can change with time as memories are consolidated into new circuits, some kind of memory is imprinted on the brain. From early development onwards, neuron interactions are constantly remoulded as synapses are made and lost in response to the continually change in incoming sensory information. When we first open our eyes, for instance, synaptic connections are refined to reinforce circuits that are used for vision. But storing memories that will help us with the most complex types of learning involves specific parts of the higher brain, notably the cerebral hemispheres.

Within the cerebral hemispheres there are dense regions of grey matter, in which cell bodies and synapses are concentrated. Unsurprisingly, the extensively folded outer layer, the cerebral cortex, plays a leading role in memory. All of our accumulated long-term memories are stored here. But other areas of grey matter also play an important role in the processing of memory. These are located deeper inside the brain and derived from the upper and lower folds of the cerebrum that project between the hemispheres.

One such structure with an important role in memory is the hippocampus, situated within each temporal (side) lobe. The hippocampus contains neurons that are responsible for generating the short-term memories that are so fleeting, but also for consolidating long-term memories as new synaptic connections are made within hours of making short-term ones. Over weeks or years, these effects then have an impact on the cerebral cortex, where synaptic connections are moulded to store memories for the very long term. By the time long-term memories are stored in the cerebral cortex, they have become independent of the hippocampus. But any new memories consigned to storage will need the hippocampus to process them.

Emotionally Charged Memories

Some memories are imprinted with more emotional charge than others, and these, too, can be explained by the physical makeup of the brain. The deepest parts of the cerebral hemispheres, close to where they join to the underlying brainstem, contain not only the memory-storing hippocampus, but other parts of the brain more intimately linked with emotion. In particular, a small knob of grey matter in front of each hippocampus, called the amygdala, is heavily implicated with emotional functions of the brain. The brain's two amygdalae are also linked to the central hormone-producing regions, such as the hypothalamus. In addition, they are closely connected to the olfactory bulb: the part of the brain that helps to process smell. This helps to explain why many smells can provoke deep emotional memories.

The entire arrangement of parts that bring together functions related to memory, emotion and smell comprises the brain's limbic system. This is described in more detail in chapter 9.

Memory Loss

Memory loss, or amnesia, can happen because of a physical injury to the brain, through disease or even as a result of a psychological trauma. There are two types: retrograde amnesia happens when memories are lost before a certain date (typically coinciding with an event, such as an accident, that caused the memory loss); anterograde amnesia occurs when short-term memory is compromised, meaning that new long-term memories cannot be created. In both cases, memory loss occurs when neurons are damaged, especially the neurons of the hippocampus. The hippocampus has the dual role of processing short-term memories and consolidating memories into long-term ones. However, long-term memories are eventually stored independently in the cerebral cortex, and this explains why amnesiacs with hippocampus damage may remember incidents that happened many years ago, while they cannot remember what happened just a few minutes ago.

Chapter 9
THE BRAIN, EMOTION AND BEHAVIOUR

Feeling Good, Feeling Bad

The brain determines our innermost feelings. Whether we are happy or sad, angry or deeply afraid, the emotions and moods we feel can be traced to the workings of neurons in the brain and the chemical signals they produce.

All of the most deeply intense and personal mental processes that go on inside our head, relating to how we feel on the inside, involve complex workings of the brain. Just like any other aspect of our behaviour, emotions and moods can be understood at the most fundamental level: the ways our brain cells are working. Not only can different kinds of emotions be associated with different parts of the brain, but they can even be linked to the balance of neurotransmitters, the chemical messengers that are produced by communicating neurons. Emotion is as much about chemistry as it is about biology.

Emotions, at their root, involve mental sensations of pleasure or distress. In one way or another they generally make us feel good or bad.

When we eat food that tastes good, we experience an emotional feeling that encourages us to repeat this behaviour.

Fear motivates us to escape from danger, while thirst drives us to seek out water. Both are feelings that are essential to survival.

Emotions have evolved as a way of helping to ensure that our behaviour is done for our best interest. Things that give us pleasurable sensations, such as food or sex, make us want to do them again. In our prehistoric ancestors this was a good thing. It helped our bodies obtain nourishment and pass on their genes. In a similar way, things that make us feel bad or fearful, such as an attack by a venomous snake, are unlikely to be repeated, so we avoid danger. For this to work, emotions must be intimately tied in with memories. And, as we shall see in this chapter, that is exactly what we see in the structure of the brain.

Primitive Feelings

Our feelings are there for a purpose. And some of our feelings are very primitive indeed. Hunger, thirst and pain are all sensations that are generated in the brains of animals to help them to survive. They involve automatic reflex responses of the kind we saw in chapter 6. But unlike the routine reflexes that are involved in things such as regulating heartbeat, the automatic trigger must also affect other parts of the brain to perform the survival strategy really well. When we feel hungry or thirsty, in both cases this begins with stimulation of a part

of the brain called the hypothalamus, the same part that is so critical to homeostasis and the flow of many hormones. The hypothalamus is in the deep unconscious region of the brain, but sends some of its signals to the "higher" parts involved in conscious decisions. Once other parts of the brain are involved, we experience a "feeling" that will help to make us to change our behaviour, although this is only part of a broader strategy for correction. For instance, the "thirst" centre of the hypothalamus works to stop our body from becoming dehydrated. If our blood water levels are low, it automatically sends signals to our kidneys to retain water. It also communicates with other parts of the brain to drive us to reach for the tap.

We wouldn't regard hunger and thirst as genuine emotions, like fear or love, but the principle of purpose is similar. Hunger and thirst are responses to basic states of the body: whether it contains enough food and water. They are concerned with homeostasis: the system that helps keep internal factors within safe limits. True emotions might instead be triggered by an outside influence, such as a dangerous snake provoking a feeling of fear, but the fact that this might help us avoid getting bitten is also good for the body!

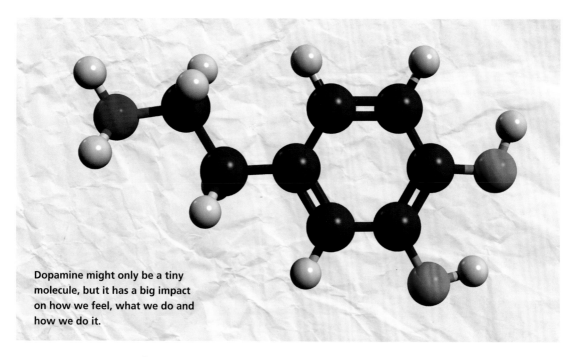

Dopamine might only be a tiny molecule, but it has a big impact on how we feel, what we do and how we do it.

The Chemistry of Feelings

The purpose of emotions is to motivate, and they do this by making us feel good or bad, depending upon the circumstance. Specifically, the good or bad feelings arise in response to a stimulus, which could be an event, a thing, another person, and so on. The good or bad feeling makes us want either to experience the stimulus again or to avoid it in the future. These pleasurable or hurtful sensations define emotions, and we learn from them. The feelings themselves are subjective, meaning that we all experience them in different ways, but they are associated with changes in the physiology of the body that can be measured. Many of these changes are controlled by the autonomic nervous system (see chapter 6). For instance, strong emotions of anger, fear or even love might result in speeding heartbeat and dilation of the pupils of the eyes: classic signs of the so-called "fight or flight" response.

Good and bad feelings are associated with different levels of the body's chemical signals: the neurotransmitters that are released between neurons, as well as the hormones that flow into the blood. This explains why a particular emotion can impact on lots of different organs at once. The neurotransmitter dopamine is produced during feelings of pleasure, whether that is from eating a cake or socializing with friends. But the stress hormone cortisol (see chapter 6) can inhibit the dopamine system, leading to opposite effects. Unsurprisingly, high cortisol and low dopamine can be associated with clinical depression.

The Brain's Pleasure Centres

Certain nuclei, mostly lying deep within the cerebral hemispheres and at the top of the brainstem, have been dubbed the brain's pleasure centres because they are implicated in the brain's dopamine pathways. In the 1950s, a pair of neuroscientists, James Olds and Peter Milner, experimented with rats and found out exactly where these pleasure centres were inside the brain. They discovered that they could artificially stimulate a pleasurable experience by administering a small electric shock through an electrode that was implanted in a particular part of the animal's brain. The shock was given every time the rat entered the corner of its box or pressed a lever. Quickly, the rats learned to associate this action with the pleasurable

The Brain, Emotion and Behaviour

feeling and went back to repeat the process again and again. The nucleus accumbens is an example of one such hedonistic hotspot, or "pleasure centre". It is located deep inside each cerebral hemisphere, in one of the so-called basal ganglia (see chapter 7) just in front of the hypothalamus. It gives us a feeling of pleasure in response to attractive stimuli. And there are other hotspots: indeed, it is thought that simultaneous activation of all of them gives us a feeling of euphoria. But the motivation involved has a complication. In addition to a "pleasure centre", the nucleus accumbens contains a "wanting centre". The two can obviously be activated together, but it is also possible for us to crave things like money that are not inherently pleasurable because we have learned to associate them with pleasurable feelings.

Research suggests that dopamine, produced by a nucleus at the top of the brainstem, controls the "wanting centre", whereas related chemical signals control of the pleasure centre. Like all neurotransmitters (and hormones), these chemicals exert their effect because they bind to particular receptors, meaning that they can affect any parts of the brain that carry these receptors. Although dopamine is the principal rewarding neurotransmitter that increases our degree of motivation, it can also activate other signallers. For instance, it may be coupled with the release of serotonin (sometimes described as the "happiness" chemical: see chapter 6), and provokes the release of noradrenalin, which enhances level of alertness. Whatever happens, the good feeling generated sticks in our minds. Just like the experimental rats with implanted electrodes, we learn to want more of the same thing. In the end, our brains are learning by a form of conditioning: see chapter 8.

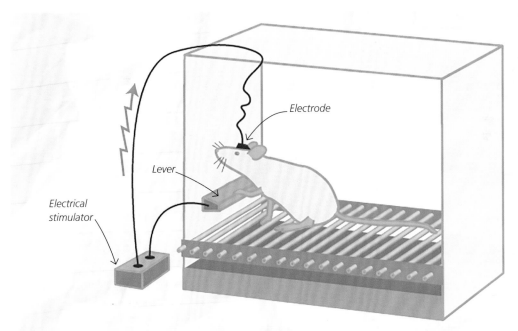

Experiments with rats suggest that they will do almost anything to get a stimulation to the "pleasure centres" of the brain. In 1954 American psychologist James Olds collaborated with Canadian neuroscientist Peter Milner to identify exactly where they were. In their experiments, rats learned that they could have their pleasures centres stimulated by pressing a lever, and they repeated the behaviour to receive the pleasurable sensation.

Cravings and Addiction

Any craving begins with an incident, in other words a stimulus that activated the brain's reward system to give a feeling of pleasure. With this feeling comes the desire to repeat the experience over and over again. When this happens, the promise of more pleasure leads to a craving that can interfere with the more rational parts of the brain, such as the prefrontal cortex. As a result, we cannot "think straight". In fact, if the body fails to satisfy the craving, it will even release stress hormones, making the desire to satisfy the craving even stronger.

Long-term changes in the brain can also make it "desensitized" to pleasure-giving stimuli, meaning that the craving actually becomes more difficult to satisfy. The brain has developed a compulsive disorder: this is addiction. At this point, the rational decision-making parts of the brain have little influence over behaviour, even when it causes harm.

Any initially pleasurable activity, such as gambling, shopping or even surfing the internet, can turn into an addiction. The effects of drug addiction are complicated by the fact that drugs themselves can interfere with the reward systems in the brain. For instance, cocaine upsets the way dopamine is produced at the synapses between neurons. Specifically, it blocks a particular protein that removes dopamine from the synapse. Normally this protein prevents the dopamine from building up, helping to give it a more measured effect. But, in the presence of cocaine, this does not happen, meaning that synapses experience an over-surge of dopamine, exaggerating the pleasurable effects, making the craving and addiction even stronger.

Gambling can turn into a compulsive disorder as the pleasure it gives becomes harder and harder to obtain. Such addictions can lead to harmful, self-destructive behaviour.

The Brain's Alarm Centres

Feelings of anger, frustration or disappointment are effectively the antithesis of pleasure. They indicate that something is wrong, and the brain needs to be alert to this fact in preparation for setting things right. In fact, these feelings involve parts of the brain that are also triggered when our sensors detect pain. The overall message is that the brain is not comfortable with the situation.

A region of the brain that might be called its "alarm centre" sits within the furrow between the two cerebral hemispheres. Specifically, it is found on a fold and groove that runs along each of the facing inner sides, called the cingulate gyrus and sulcus. These parts receive signals coming from pain receptors around the body and, in response, provoke the automatic symptoms associated with pain, such as raised heart rate, increased stress and contorted facial expression. But they are also active when we feel distressed, even when there is no physical injury. The cingulate gyri are therefore used by the brain to focus our attention on an emotionally distressing matter that demands special attention. Even momentary surprises, including nice ones, will trigger these alarm centres.

A related region in each hemisphere, the amygdala, is part of this complex of structures (called the limbic system: see next section). The amygdala is also part of the alarm system, since it can help activate the adrenaline rush that is part of the body's "fight or flight" response.

The "alarm centre" in the brain sits between the two cerebral hemispheres.

Expressing Emotions

Although the body seems to swing between sensations or pleasure and discomfort, there is obviously a much broader range of human emotions that we can experience. English naturalist Charles Darwin thought that our emotions were products of evolution by natural selection, and that there were equivalent feelings in animals. In 1872, he published his thoughts on the matter in *The Expression of the Emotions in Man and Animals*. Darwin believed that the same kinds of emotions were universal in the natural world, and that they were expressed in consistent ways across humanity.

For Darwin, expressions of emotions such as joy and anger, as seen on human faces, could be recognized all over the world, even in different cultures. Some anthropologists have since argued that facial expressions were learned in different ways in different cultures, but a century after Darwin proposed it, American psychologist Paul Ekmon took to testing his idea more objectively. Using pictures of actors feigning emotions, Ekmon showed that the many of the same expressive emotions could be recognized all around the world, including in tribal peoples in New Guinea who had been isolated from the wider world. Today, Ekmon's system for characterizing facial expressions with emotions is used to help people with social communication disorders such as autism. In 1980, American psychologist Robert Plutchik classified and quantified emotions not only according to whether they are pleasurable or discomforting, but also according to their strength, or level or arousal. Plutchik's resulting "wheel of emotions" remains an important way of categorizing human emotion.

Sadness

Happiness

Surprise

Disgust

Anger

Fear

In 1976 Paul Ekmon compiled a collection of images that simulated human emotion, called "Pictures of Facial Affect". This and similar resources have been used to test the idea that facial expressions might be universally recognized the world over.

The facial expression system is an automatic, and honest, advertiser of our mood. But it is possible to override it. If the conscious parts of the brain have time to anticipate when this is desirable, they can control the facial muscles in a voluntary way. Such a "poker face" might be used when faked facial expressions are called upon to deceive others.

Empathy

The very fact that humans can communicate the way they feel by their facial expressions adds an interesting dimension to the biology of emotion. Advertising emotions is important for social interactions. Darwin believed that emotions had evolved for a purpose: by reading the facial expressions of others, we develop a sense of empathy. Neuroscientists have discovered neurons in the front of the cerebral cortex, the prefrontal cortex, that mimic the activities that we see in others. Called mirror neurons, they fire both when performing an action and when seeing the same action performed by another individual (and even

The Brain, Emotion and Behaviour

when thinking about the action). They are thought to help with learning through copying the action of others, especially in children. But some neuroscientists think that mirror neurons are also important in determining our sense of empathy. By recognizing and empathizing with expressions and behaviour in others, our prefrontal cortex can react appropriately to social signs. This may be important in the way we develop a sense of fair play, repressing selfish behaviour and helping us to relate to others. In other words, this region of the brain is implicated in moral decision-making. People who suffer from injuries to the prefrontal cortex may be left with impaired moral judgement, and sometimes develop psychopathic tendencies.

Specific chemical signals have been linked to empathy. The hormones vasopressin and oxytocin have been shown to be important in helping people "read" emotions. Abnormal levels of both these hormones have been found in people with autism.

The Cause of Emotions

Despite the wealth of studies supporting the idea that emotions depend on the circuitry and chemistry of the brain, the root cause of emotion remains a contentious point. One theory has it that emotions are always products of the body's physiology. In other words, what we feel on the inside stems from the physical workings of our

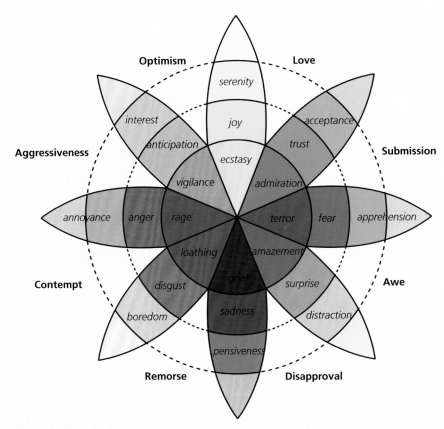

Paul Plutchik's Wheel of Emotion was devised as a way to classify emotions according to their strength (the level of arousal) and valency (the degree to which they were associated with pleasure or displeasure).

The ways in which we express our emotions vary depending on our personalities, which may be introverted or extroverted.

body: for instance, how the autonomic nervous system drives our heart rate, and how our surging hormones and neurotransmitters change the way our cells work. But other biologists think that physiology alone could not explain the speed and variety of emotional surges. Experiments have shown that different emotional states can be associated with the same kind of physiology. Perhaps stimuli themselves trigger emotions in the brain directly.

Unsurprisingly, there is a genetic impact on our emotions. Some people carry genetic variations that cause the brain's emotional systems to be disrupted, meaning that, for instance, they have difficulty reading the emotions of others. And, as usual with any characteristic, the environment and the person's upbringing also exert important influences. Overall, this means that people can differ a great deal in the ways they outwardly express their emotions depending on whether they are extroverted or introverted. Although the mechanism for expressing emotions may be much the same, their intensity, and the triggers that cause them, are not completely consistent. Indeed, they help to account for our different personalities.

Mood-changing Drugs

Given that so much of the emotional system of the brain is based on chemistry, it should come as no surprise that many drugs can affect this system to alter our emotional state. Drugs that do so typically work by interfering with the chemical signalling by neurotransmitters in synapses: the gaps between the neurons. These neurotransmitters can have such pervasive effects in different parts of the brain that drugs that block them or intensify them end up having multiple effects. For instance, LSD (lysergic acid diethylamide) probably prevents serotonin, the "happy" signal, from being removed from synapses, so enhancing its effects on the rest of the brain. Serotonin not only determines mood, but is also involved in processing sensory information in the cerebral cortex. LSD therefore can generate hallucinations, and its effects on the mood system mean that some of these "trips" are pleasurable, but others are negative, causing anxiety, paranoia or even potentially suicidal thoughts.

The Secret Emotional Brain

Emotions are controlled by regions of the brain that lie deep inside the cerebral hemispheres. This collection of structures, called the limbic system, determine our deepest feelings, and has been referred to as the "visceral brain".

In the 1950s, American neuroscientist Paul MacLean became interested in how the brain was involved in controlling emotion, and began to seek an evolutionary explanation for the emergence of the "emotional brain". His idea was that the higher parts of the brain (in other words, the cerebral hemispheres) could be divided into three regions that corresponded with different stages in the evolution of vertebrates. The "reptilian" brain was the lowest region involved in instinctive aggression, territoriality and so on. The "old mammalian" brain, on the inner parts of the hemispheres was concerned with emotion. The "new mammalian" brain was the cerebral cortex around the outside, involved in the highest functions, such as planning, language and abstract thought. We now know that MacLean's lowly "reptilian" and emotional "old mammalian" brains actually occur across all backboned animals, so probably appeared early on with the first fishes, and it seems that even animals such as sharks have emotions. The cerebral cortex probably originated in the first mammals, but it is arranged more simply (without the folds) in "primitive" mammals, such as the platypus. But MacLean's so-called triune model still helps to put the layers of the brain in some sort of evolutionary context, including the "emotional brain" that is sandwiched in the middle.

MacLean's emotional brain had already been described a century earlier. The French physician Paul Broca (who is more famous for discovering a language-controlling area of the brain: see chapter 7) had observed that this part of the brain seemed structurally different from the rest of the cerebral cortex. He called it the "limbic" system, from the Latin, limbus, meaning border, a reference to the fact that it was mostly positioned in the border between the cerebral hemispheres and the brainstem. Today, some neuroscientists think that the term should be abandoned because this entire region is really made up of a number of disparate parts of the brain that are linked only by function. And, as we shall see, some of its parts are concerned with seemingly unrelated tasks, such as memory and sensing smell. But the overall association means that an "emotional brain", whether you call it the limbic system or not, certainly exists. Meanwhile, functionally equivalent brain structures have also evolved in invertebrates to drive similar behaviours.

Some mammals, such as the egg-laying platypus, have a brain that retains the smooth, unfolded cerebral cortex of mammals' reptilian ancestors.

Where is the Emotional Brain?

The complex interconnections of the limbic system give us a clue as to how it affects behaviour. Its most primitive components are a pair of smell-sensing olfactory bulbs that lie underneath the cerebral hemispheres. As we have seen (chapter 7), these were probably prominent in the earliest backboned animals, suggesting that the sense of smell was important from an early stage in evolution. Today, the olfactory bulbs are enlarged in many vertebrates, from sharks to mammals. They connect to a loop of grey matter that encircles the fluid-filled ventricle (chamber) in each hemisphere.

This combines parts concerned with receiving other sensory signals (the thalamus), storing memory (the hippocampus), and imprinting emotional charges on these memories (the amygdala). It also communicates with the core areas of the brain involved in homeostasis and hormonal regulation (such as the hypothalamus). Together, the olfactory bulbs, loops of grey matter and these regulatory parts of the brain make up the limbic system.

The limbic system, therefore, enables impulses related to smell, emotion, memory and regulation to combine and interact. In addition, the system has connections with the cerebral cortex on the

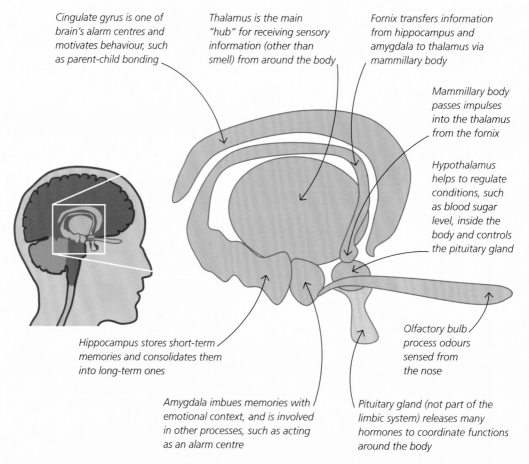

Cingulate gyrus is one of brain's alarm centres and motivates behaviour, such as parent-child bonding

Thalamus is the main "hub" for receiving sensory information (other than smell) from around the body

Fornix transfers information from hippocampus and amygdala to thalamus via mammillary body

Mammillary body passes impulses into the thalamus from the fornix

Hypothalamus helps to regulate conditions, such as blood sugar level, inside the body and controls the pituitary gland

Olfactory bulb process odours sensed from the nose

Hippocampus stores short-term memories and consolidates them into long-term ones

Amygdala imbues memories with emotional context, and is involved in other processes, such as acting as an alarm centre

Pituitary gland (not part of the limbic system) releases many hormones to coordinate functions around the body

The limbic system is a complex arrangement of structures that are involved in emotion and memory. All of the structures shown here, except the hypothalamus and pituitary gland, are paired in the cerebral hemispheres.

The Brain, Emotion and Behaviour

Many mammals, such as dogs and rabbits, have a highly developed sense of smell. They have large olfactory bulbs to process information from their noses. The bulbs form part of the limbic system, meaning that smells are likely to produce strong emotional responses in these species.

surface of the cerebral hemispheres, which controls complex conscious thought. This means that our deepest feelings, related to emotion, hormone levels and strong memories, are seated in the limbic system. The entire arrangement is strongly influenced by a sense of smell, explaining how smells can often conjure up some of the most emotive memories of all. In general terms, the limbic system provokes arousal in relation to feelings that are pleasant or unpleasant, some of which are recalled from memories.

The Origins and Purpose of Emotion

If the limbic system first evolved in our primitive fishy ancestors, does that mean that anything from guppies to geckos, ducks to dogs, experiences the same emotions as us? Probably not. In most of our backboned ancestors, the limbic system was first dominated by the need to process the sense of smell, and it can still provoke a strong reaction in those with highly developed olfactory bulbs. A shark is driven into a feeding frenzy by sensing blood in the water, while a mammal that smells a pheromone, the perfume given off by a possible mate, might become sexually aroused. If a sense of smell was originally so important for finding food or sex, then it is not surprising that it evolved into a system that was so powerfully linked to sensations of pleasure or disgust. It is even connected to our stored memories and to deeper, more primitive, functions for controlling the body's interior, such as those involved in releasing hormones.

Of course, our own sense of smell has now diminished (although we retain many things that go with it, such as "smell" genes and a diversity of receptors: see chapter 7). Other senses such as vision have assumed greater importance, while our innermost limbic system, generating the emotions and emotional memories, remains central.

Ground Plan of the Limbic System

The brain's smell system begins when scents are detected by receptors in the nose, which then trigger impulses along the olfactory bulbs that lie underneath the cerebral cortex. As we saw in chapter 7, the olfactory bulbs help to process odours, just as a part in the core of the brain (the thalamus) acts as a preliminary processing centre for other senses. But the olfactory bulb is also linked to higher parts of the brain involved in arousal and attention.

The almond-shaped amygdala sits deep within the medial temporal lobe. It is connected to many different parts of the brain, and helps to give our memories an emotional charge.

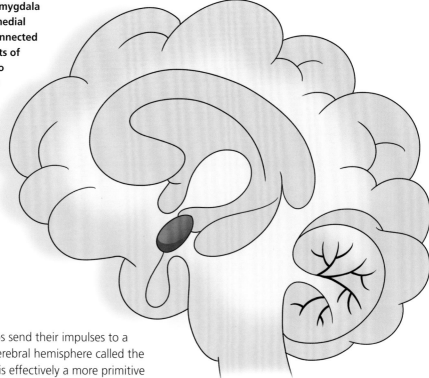

The olfactory bulbs send their impulses to a lower part of each cerebral hemisphere called the piriform cortex. This is effectively a more primitive part of the cerebral cortex that continues with processing of the sense of smell. But it also acts as the link to other parts of the brain that enable smell to combine with other functions of the limbic system: the amygdala and hypothalamus. The amygdala is named from the Greek for "almond", a reference to its shape. It plays a central role in processes of the limbic system: in the brain of any backboned animal, the amygdala links the smell-sensing system with memory. It does this by passing information on to its neighbour, the hippocampus. This is the part of the brain that stores and processes short-term memories. The amygdala therefore helps the brain to learn from the experience of sensing odours. This means that odours can serve to reinforce or inhibit certain kinds of behaviour. Through all these connections within the limbic system, the amygdala helps to imprint certain memories with an emotional charge. The olfactory link to the hypothalamus can provide even stronger influences on behaviour. Since the hypothalamus is involved in the release of hormones, including those that control sexual activity, the olfactory parts of the limbic system may specifically influence aggressiveness and mating behaviour.

Such is the ground plan of this section of the limbic system for most backboned animals. But in humans, at least, the paired amygdalae seems to do far more. They have been called centres of "fear and aggression", but this is an oversimplification. For instance, the amygdalae found in the left and right hemispheres seem to do different things. If the left amygdala is artificially stimulated, it might provoke emotions of joy or distress, but stimulation of the right one always produces negative responses. There are sex-specific differences too: the amygdalae are proportionately bigger in men than women.

The Limbic System at the Core of the Higher Brain

The "core" of the higher brain, situated immediately above the brainstem, is involved in coordinating sensory input, as well as regulating

The Brain, Emotion and Behaviour

vital functions (see chapter 5). In particular, it contains the sensory hub (the thalamus) and the regulation centre linked to the hormonal system (the hypothalamus). And parts of this core are components of the limbic system. Here, some of the brain's nuclei serve as relay centres for passing information between parts such as the amygdala, hippocampus and thalamus, or to send information to and from the higher parts of the brain and the brainstem. Other centres are concerned with regulating the neurotransmitters dopamine and serotonin, which help to regulate mood. The front limbic section of the thalamus is also involved in controlling level of alertness and spatial navigation. Finally, as we have seen above, the hypothalamus helps to control the body's regulatory systems and hormones, some of which can heavily influence mood and behaviour.

Limbic Components in the Cerebral Cortex

The highest parts of the brain also contain parts of the limbic system. These are found in the cerebral cortex within the deep groove between the two cerebral hemispheres. They are dominated by a long, arched fold that runs down the inner surface of each hemisphere: the cingulate gyrus. This fold receives impulses from the thalamus underneath, as well as the cerebral cortex above. In the previous section, we saw how the cingulate gyrus acts as part of the brain's alarm system, by intercepting pain signals and inducing the appropriate physical and emotional responses. It is also involved in motivation, helping to provoke a repeat of behaviours associated with pleasurable feelings. It is active in parents, both mothers and fathers, and seems to help them bond with their new arrivals.

The cingulate gyrus in the limbic system is activated in new parents, and this help them to bond with their baby. Since this part of the brain is also involved in motivation, it also helps to spur the parents on to make an extra effort to provide for the child.

A Limbic System for a Social Animal

All animals, including humans, normally behave in ways that increase the chances of survival. As we have seen throughout this book, the brain controls behaviour by reacting to sensory cues and combining this information with memories stored in the connections between brain cells. Humans are a highly social species, so many of these cues come from members of our own kind. Just like other social animals, we find mates, rear families, defend our place and recognize our position within society. This involves picking up on very complex social cues, such as reading facial expressions or being drawn to a crying baby, and reacting appropriately. The brain's limbic system is at the heart of this control. These behaviours are also innate, meaning that we can be driven to perform them automatically. In that respect, the limbic system is like an extension of parts hidden deeper in the brainstem that control vital functions, such as heartbeat and body temperature.

Many innate aspects of our behaviour, such as fear of predators, the bonding between parent and child and sexual attraction, are "hard-wired" into the limbic system: we do not have to learn about them for them to work. But the limbic system's pleasure and alarm centres also help us to use learning to reinforce some behaviours and avoid others. As we shall see in chapters 11 and 12, the secrets of the limbic system may explain some differences between the behaviour of the sexes, and how our behaviour changes as we grow up.

At any social gathering, we are constantly monitoring the social cues of others and deciding how to respond appropriately. The limbic system allows us to do much of this without conscious effort.

The Brain, Emotion and Behaviour

Chapter 10
THE INTELLIGENT BRAIN

Evolution of the Brain

Humans emerged from ape-like ancestors nearly ten million years ago, and our brains have been evolving ever since. As we evolved into modern humans, our brains became bigger to cope with complicated thoughts and complex skills.

Humans are primates, a group of mammals with grasping hands and big brains. Specifically, we are apes that stand upright and have practically hairless skin over most of our bodies. But perhaps it is our brains that mark us out most of all. For years, biologists have looked deep into our brains and compared them with our closest relatives in an attempt to find out exactly what the differences might be. But the structure of the human brain, with its folded cerebral cortex, brainstem and cerebellum at the back, is built with the same organization as that of any other primate, and indeed any other mammal. And, as we have seen, if we reach out to our more distant relatives, we find there is a ground plan for brain structure that is the same across all animals with a backbone.

Humans are a highly social species, and we live together in huge, complex societies. We need big brains to enable us to navigate our way through life.

This should come as no surprise given that all life on earth has descended from a single common ancestor. As new life forms evolve, they end up with differences in anatomy and behaviour, but all life will always be connected, in one way or another, because we all have attributes that we have inherited from our ancestors, and the brain is no exception. Today, biologists can use a wealth of techniques to study animal behaviour, animal brains and even the fossilized remains of creatures that lived in the past. The combined results of these studies give us a better understanding of how the human brain came to be and how our intelligence evolved.

Origins of the Primate Brain

The earliest fossils of primates are creatures that might have resembled modern-day bushbabies, and lived some time after the demise of the dinosaurs 65 million years ago. But given the fragmented nature of the fossil record, it is likely that the very first primates were even older than this, and may have shared their world with the great reptilian beasts. These animals would have been tree-living creatures, with brains that were probably not much bigger than those of other similar-sized mammals. They shared with other mammals a key feature of the brain that would be critical to future developments. A region of the cerebral cortex called the dorsal pallium was bigger and more extensive in early mammals than in their reptilian ancestors. In mammals, it evolved into the cerebral cortex: the outer layers of the brain that are so important in "higher-level" thinking and problem-solving. And for many mammals, including primates, this cerebral cortex expanded so much

Living primates are split into two groups. The prosimians (meaning "before monkey") include mainly nocturnal bushbabies (above) and lemurs. They have brains that are good at processing odours, like those of early mammals. The simians, including monkeys (right), apes and humans, have bigger brains for daytime activity.

that it became folded and grooved beneath the confines of the skull. The evolution of an expanded, convoluted cerebral cortex explains the intelligence seen in mammals today, although some still retain smooth cerebral hemispheres (such as in the platypus, opossums and many rodents).

Today, there are primates that haven't changed much from those ancient tree-dwelling ancestors. Lemurs and bushbabies have the folded brains, but they sense the world in a different way from us, relying heavily on smell. Most are nocturnal, like many other mammals, and an advanced olfactory system is the best way to navigate and socialize in the dark. But about 30 million years ago, a new

kind of primate appeared with a new strategy for living. The first monkeys differed from lemurs in many physical ways: they had nails instead of claws, and flatter faces. But they also behaved differently. They moved around during the day and began to rely more on other sensory systems, particularly vision and touch (especially with their extra-sensitive fingertip pads). As a result, their olfactory systems diminished until they were just sufficient for detecting odours. This meant that some areas of the cerebral cortex, including those for vision and tactile sensations, expanded. And with them evolved the areas involved in complex decision-making and reasoning.

Neanderthals lived in Europe until around 40,000 years ago. Today, they are considered by some to be a subspecies of our species _Homo sapiens_. They had even larger brains than us.

Brain Capacity

We can calculate the size of the brains of fossil primates from the sizes of their skulls. Bigger skulls accommodate bigger brains. Measuring the capacity, or volume, of the skull gives a good indication of the size of the brain that was once inside it. Today, the smallest living true apes (aside from humans) are the chimpanzee and orangutan. The chimp cranium has a volume of around 400 cubic centimetres (25 cubic inches), while that of the orangutan is a little smaller. In gorillas, unsurprisingly, the brain is larger at around 470 cubic centimetres (29 cubic inches), but that has happened roughly in proportion to the increased size of their bodies. Bigger bodies need bigger brains to keep them working. This means that the animal with the biggest brain of all is not a primate at all:

the sperm whale's brain measures in at a whopping 8,000 cubic centimetres (500 cubic inches). But among primates, when we come to humans, there is an enormous increase in brain size. Even though gorillas are more than three times our body weight, the average cranial capacity of a modern adult human is 1,260 cubic centimetres (77 cubic inches). That puts us far ahead in the brain stakes.

The enlargement of the human brain appeared to start when prehistoric ancestors descended from the trees, probably as a result of habitat change that drove them to exploit more open grassland habitats. By 2 million years ago, one of the first recognizably human species, _Homo habilis_, had a cranial capacity of 700 cubic centimetres (37 cubic inches), significantly larger than that of modern chimpanzees. This rose to more than 1,000 cubic centimetres (60 cubic inches) in a later species _Homo erectus_, and peaked in Neanderthals within the last half a million years at up to 1,900 cubic centimetres (116 cubic inches). This means that Neanderthal brains were even bigger than those of humans living today. However, they also had larger bodies than us, and in proportion to their body size, their brains were about the same as ours.

A continuing expansion of the cerebral cortex accounted for the increase in human brain size, but some areas expanded more than others. The biggest expansions happened in the front part of the frontal lobe and in the temporal lobes at the side. The front of the cerebral hemispheres, the prefrontal cortex, is involved in some of the most complex thought processes, such as reading the feelings of others and the sense of right and wrong; these are things that are perhaps most close to what makes us human (see next section). The temporal areas, in contrast, are involved in very specific skills: language processing and language cognition (see chapter 7). As the entire brain grew bigger, this was at the expense of some other physical parts of the head. In particular, it meant that there was less room for big muscles to work heavy jaws. Our jaws became smaller and our faces grew flatter.

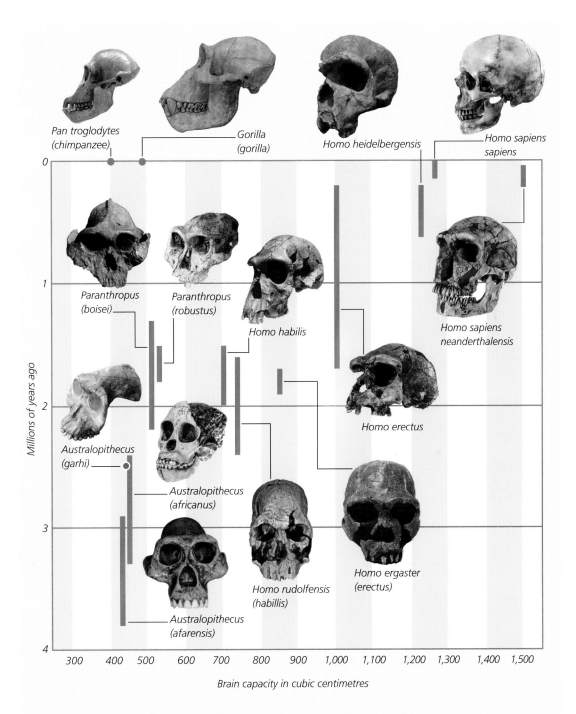

Pan troglodytes
(chimpanzee)

Gorilla
(gorilla)

Homo heidelbergensis

Homo sapiens
sapiens

0

1

Paranthropus
(boisei)

Paranthropus
(robustus)

Homo habilis

Homo sapiens
neanderthalensis

Millions of years ago

2

Australopithecus
(garhi)

Australopithecus
(africanus)

Homo erectus

3

Homo rudolfensis
(habillis)

Homo ergaster
(erectus)

Australopithecus
(afarensis)

4

300   400   500   600   700   800   900   1,000   1,100   1,200   1,300   1,400   1,500

Brain capacity in cubic centimetres

Cranial capacity can be measured for all living apes, as well as for extinct species where there is sufficient fossil material available. The results show that average brain size has increased significantly in the evolution of modern humans.

Larger cerebral hemispheres allowed our human ancestors to cooperate in complex social groups.

Why Did Our Brains Get Bigger?

There is no single explanation for why our brains evolved to be bigger, but our search for clues must begin with the shift in lifestyle that doubtless started the process off: when walking humans evolved from climbing apes. The ability to stand upright on two legs, called bipedalism, was actually emerging long before brain size started to grow, and many factors have been attributed to the literal rise of walking humans. As forests gave way to savanna, habitat change may have forced our ancestors to abandon their climbing ways, but other advantages of walking on two legs would have been passed on by natural selection. Upright walking and running may have been faster for escaping predators across open land. And by freeing up the grasping hands, upright humans could collect food more efficiently. This probably led to better manipulative skills too, opening the way for making tools. But whatever the reason, bipedalism placed certain demands on the brain. Excellent coordination skills are needed to balance

the upright body to stop it from falling over. This demanded a bigger cerebellum: the part of the brain involved in automatic control of fine motor skills (see chapter 5).

But, as we have seen, the biggest development of all happened in the cerebral hemispheres, and the fact that this was concentrated in the prefrontal cortex and temporal lobes is highly suggestive: these are the areas tied up with social skills and language. Humans, like most primates, are highly social animals. There are benefits of living in groups: it offers better protection from dangerous predators and can make foraging or hunting for food more efficient. And these benefits emerge more clearly if members of the group cooperate with one another. For that to happen, there needs to be a good system of communication. Members of the group have to use signals of communication, and be able to recognize them when used by others. Parts of the prefrontal cortex are used to "read" the thoughts of others, helping us to sympathize and empathize (see chapter 9), while spoken language made communication more precise and less ambiguous. Interestingly, the enlarged cerebellum, already involved in controlling upright gait, may have helped further in coordinating the complex movements involved in speech.

A Painful Birth

No-one would argue against the idea that a bigger brain is an advantage for a complex social primate, but as with all traits, there are biological constraints that place limits. A bigger brain inside a bigger head means that mothers go through a lot of pain when they give birth. And interestingly, this in itself can be linked to a whole range of other characteristics that make us distinctly human.

Giving birth is doubly problematic for a big-headed primate that happens to walk on two legs. One of the many adaptations of walking upright involves a curved spine sitting above a smaller pelvis with a narrow pelvic canal. Although women have wider hips than men in preparation for childbirth, they still have a much tougher time of

The Intelligent Brain

it compared with primates with smaller brains. In fact, brain size in humans grows so big that their babies must be born sooner with underdeveloped brains for childbirth to be safe. The brains of newborn humans are only 25 per cent fully developed. In our closest relatives they are up to 50 per cent developed.

These aspects of basic human biology almost certainly had a big impact on social behaviour by strengthening the bonds between members of a social group. Mothers giving birth relied more on the assistance of other members of the group. (Related apes, in contrast, may give birth quietly and alone.) And a helpless, underdeveloped newborn needs more intensive care over a longer period of time. This demands stronger bonding between the parents to get their offspring through childhood. (This may account for the fact that a woman's sexual cycle has no obvious signal to show when she is fertile. In other primates, like most mammals, there are visual signs to show males the best time to mate. Human fathers are perhaps more likely to stay around to help if they are kept guessing.) The social demands placed on a group with birthing mothers reinforced the very parts of the brain that made them big in the first place. Humans needed to be brainy not only to live successfully in a group, but also to breed successfully.

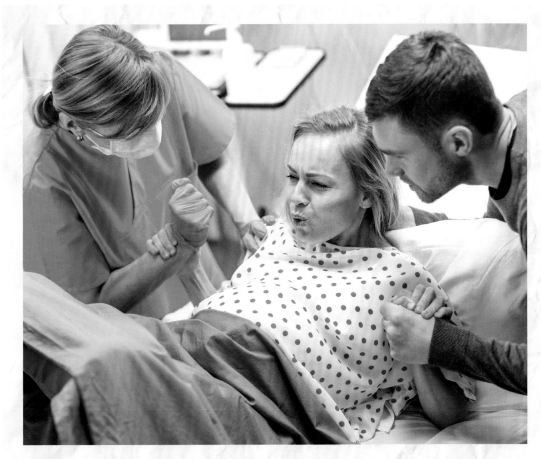

Giving birth is a difficult and painful process for a human mother. Unlike smaller-brained apes that give birth alone, humans need the assistance of others to have a successful birth.

What is Intelligence?

An intelligent brain can cope with behaviours that go far beyond automatic control of things like heartbeat. It can solve problems, make decisions and even be tuned in to the thoughts of others.

It has been said that if the brain were so simple that we could understand it, then we would have to be so simple that we couldn't. But that doesn't stop scientists trying. Understanding the most complex higher parts of the brain, both in terms of structure and function, undoubtedly represents one of the biggest challenges for biology. Scientists are getting closer to understanding how the "higher" brain works in terms of the behaviour of individual neurons and their connections. However, many of the highest brain functions, such as reasoning, morality and emotion, are emergent properties: they are more than the sum of the brain's parts.

There is no single definition of intelligence. Rather, there are a number of different definitions, and we should understand the capabilities of a

Chess is a game that requires planning and the ability to form strategies.

"clever" brain more in terms of its individual abilities. Overall, intelligence is a measure of how the brain uses or processes information, both information continually received from our senses and information stored in our memories. It helps to look at the behaviour of other animals to appreciate how humans may be superior in this respect. Most importantly, intelligence involves the kinds of higher-level skills that take behaviour away from simple stimulus–response activities and routine learning, such as habituation and conditioning. However it is defined, intelligence helps individuals to have insight into what they are doing: to plan ahead and use some kind of reasoning in the thought processes.

Encephalization Quotient

If intelligence is simply a measure of using and storing information, then all animals are intelligent to varying degrees. As we have seen (chapters 1 and 8), even the simplest nervous system can process information by detecting stimuli and sending signals to muscles that bring about a responsive movement. The flow of these signals is a flow of information. And any nervous system can store memories (again, information) as new connections are made between the neurons at their synapses. As brains get bigger, and the number of neurons multiplies, more information can flow, and more can be stored.

But clearly, our usual concept of intelligence goes much further than this: it depends upon how the information is used. More complex animals have evolved more complex "higher" parts of the brain – the cerebral hemispheres – to do just that. Animals with proportionately larger cerebral

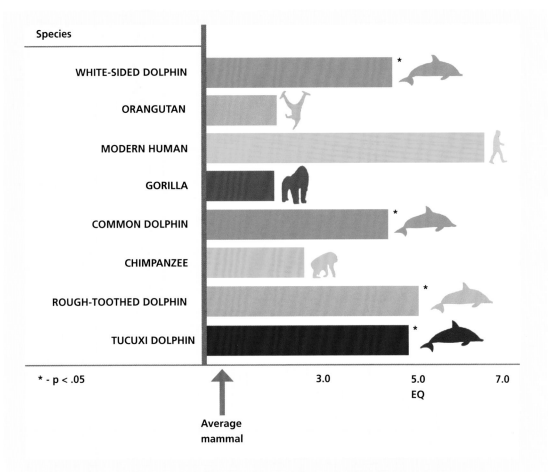

Species

WHITE-SIDED DOLPHIN	*
ORANGUTAN	
MODERN HUMAN	
GORILLA	
COMMON DOLPHIN	*
CHIMPANZEE	
ROUGH-TOOTHED DOLPHIN	*
TUCUXI DOLPHIN	*

* - p < .05

3.0   5.0   7.0
EQ

**Average
mammal**

Encephalization quotients (EQs) are measures of brain sizes that exclude the effects of overall body size and the how brains devote part of their function to routine regulation. EQs could be the best indicator we have to compare levels of intelligence among different mammals.

hemispheres are usually perceived as being more intelligent and, in this respect, humans are just about the most intelligent of the lot.

But can brain size really be used as a basic indicator of intelligence? As we saw in the previous section, scientists studying human evolution have relied on comparing cranial capacity of modern humans and fossil species to try to understand how "higher-order" brain functions have evolved. But brain size alone can be misleading. Although bigger bodies need bigger areas of brain to sense and control the different body parts, the automatic control of vital functions such as heartbeat and breathing needs about the same number of neurons, no matter what the animal.

Neuroscientists have devised a measurement called the encephalization quotient (EQ) to overcome this issue. EQ is a measure of the development of the "higher" parts of the brain only (specifically for mammals): the ones that are not concerned with vital regulations. It is an indicator, therefore, of the amount of brain that is used for more complicated mental tasks, and is often used as a guide of intelligence.

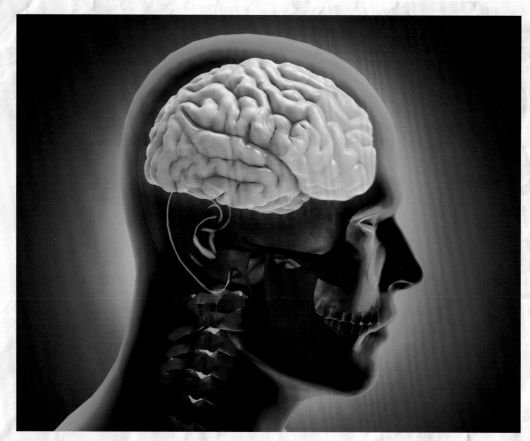

The prefrontal cortex includes the front part of the brain's frontal lobes, and covers a region of cerebral cortex that corresponds with 14 of the so-called Brodmann areas (see chapter 7).

The Importance of the Prefrontal Cortex

Right at the front of the brain's cerebral hemispheres, covering much of the frontal lobe, is a region of the brain called the prefrontal cortex (PFC). It is described as the controller of "executive" functions of the brain. This means that it is the overall monitor and arbiter of what we do: it helps us choose the best way to achieve our goals. If the brain has a "seat of intelligence", the prefrontal cortex could well be it.

This area of the brain is not just involved in problem-solving. It is strongly influenced by other parts of the brain involved in our emotions and memory, such as the limbic system (see chapter 9).

As a result, it takes cues from how we are feeling, as well as reading the feelings of others. The prefrontal cortex helps us to sort through conflict: it helps us to distinguish good from bad, both in terms of what is best for us and what is best for other members of our social group. This means that it is also used to help us behave in a way that is appropriate within a social setting: to encourage things that would be socially acceptable, and inhibit things that are not. Central to this is our ability to think in abstract ways.

Self-awareness may also be controlled by the prefrontal cortex, and whether animals are self-aware has sometimes been used as an indication of a higher-level functioning akin to intelligence.

The Intelligent Brain

Self-awareness is clearly useful in a social setting, where goals and priorities of "self" need to be balanced with the perceived interests of others. The mirror test was devised by American psychologist Gordon Gallup Jr in 1970 in an attempt to identify animals that may be self-aware. An animal is marked on a part of its body, then presented with a mirror. If the animal touches the mark on its body, this indicates that it is recognizing itself in the reflection, rather than another individual. The animals that have passed the mirror test so far include chimpanzees, bonobos, orangutans, dolphins, elephants and magpies. Pigeons may also pass the test. Most animals fail the test, but this could be explained by other factors, such as the fact that many animals rely more on sensory cues other than vision. For instance, dogs fail the mirror test, but dogs rely far more on their sense of smell when identifying individuals than their sense of vision. In human children, a positive result is typically indicated at the age of about 18 months (see chapter 12).

Components of Intellect

In 1925, British psychologist Charles Spearman used statistical techniques to devise a general intelligence factor, which became known simply as the "g factor". He looked for correlations in the performance of different people on intelligence tests: ways that their assessment of their cognitive abilities might be similar. Although the g factor is not the final all-encompassing universal measure of intelligence, it seems to be a good indicator of the potential intelligence of any one individual, and is said to account for around 50 per cent of intelligence test scores. Moreover, an analysis of the g factor can also be used to assess intelligence in animals, helping to compare their cognitive performance with that of humans.

Elephants are a species that passes the mirror test. This means that they understand that the image they look at in the mirror is an image of themselves.

Scientists have attempted to devise ways of quantifying or characterizing intelligence specifically for humans. In 1963, British-born psychologist Raymond Cattell classified human intelligence into two facets by differentiating between knowledge and problem-solving. He called these crystallized and fluid intelligences, respectively. Crystallized intelligence represents the accumulation of intellect through life, as knowledge is built up and skills are mastered. It relies on information that is stored in the long-term memory, as well as on skills that develop through experience. General knowledge and vocabulary are examples of crystallized intelligence. It increases with age as more knowledge is accumulated over time. Fluid intelligence involves novel problem-solving and reasoning without any recourse to knowledge accumulated from the past. It also involves the brain's ability to recognize patterns. In other words, it is a measure of the brain's ability to work things out based on unfamiliar circumstances. Fluid intelligence involves a good working (short-term) memory, and people with a high fluid intelligence generally go on to develop a high crystallized one.

Intelligence Quotient

Most intelligence tests are designed to take both crystallized and fluid intelligence into account, but no test can be considered the final decider of someone's intelligence level.

Best known is the intelligence quotient, or IQ, test, which is based on the work of the German psychologist William Stern in 1912. Contrary to popular belief, IQ is not an objective measure of intelligence. Rather, it is an indication of an individual's relative intelligence compared to the rest of the population. The average IQ is set to be 100, with values lower indicating below-average intelligence, and higher being above-average. Two-thirds of a population will have an IQ score from 85 to 115. The fact that IQ is a relative score in this way means that values are not comparable between populations. In other words, individuals from two populations could both have IQs that are average for their population (both would score 100), but one would be smarter than the other if, overall, one population was smarter than the other. Indeed, within populations that have been given IQ tests over many decades, such as the United Kingdom, the scores need to be recalibrated from time to time as generally it has been found that people's performance has improved. Kids may really be smarter than their parents!

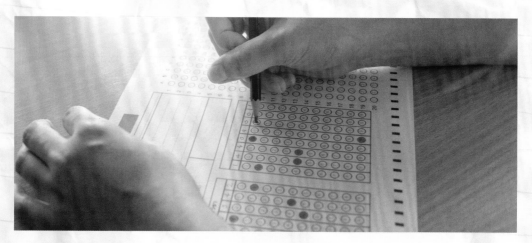

The IQ test includes a range of questions to measure the ability to reason. It has been criticized by some scientists who say that something as complex as intelligence cannot be reduced to a single number.

The Intelligent Brain

Chapter 11
THE BRAIN, SEX AND GENDER

Female Brain, Male Brain?

Differences between the sexes are ultimately controlled by genes and chromosomes, and as the brain develops, it helps to shape the bodies of girls and boys.

We all begin our lives as embryos with no sex differences. In the first few weeks of growth, the embryo is concentrating on forming the basic body parts: head, body, limb buds and organs. Its sexual organs are not yet determined, and it is said to be "bipotential". Along with all the other internal organs, sex organs start to form but they are ambiguous. It is only when the embryo reaches its sixth week, when it is scarcely bigger than a blueberry, that the first external signs of sex appear.

But its sex was determined much earlier than this, at the very point that a sperm fertilized an egg. This is because sex is written in the genes and chromosomes. Each of our body cells contains 46 chromosomes in total: the tiny threads of DNA that carry all the information needed to build a body. Two of these chromosomes are the sex chromosomes. If they are of the same kind – if they are both X chromosomes – the result is a girl. If they are different kinds – one an X, the other a Y – they produce a boy. At the same time, many more sections of DNA on lots of different chromosomes provide the instructions needed to build a human brain. And because the cells of the embryo are following these instructions far sooner in the plan of development, the brain itself appears sexless at first, just like the rest of the embryo.

Week six is the critical point as far as the sex of the developing embryo is concerned. That's when the genes on the sex chromosomes start to show their effects. They cause the primordial sex organs to differentiate as they develop. In females they become ovaries, while in males they turn into testes. From then on, a sequence of events work on the growing foetus like a domino effect as the ovaries produce female hormones and testes produce male ones. These sex hormones become important triggers in fixing the development of other primary sexual characteristics: the physical sex-specific parts that are visible when the baby is born. The hormones that circulate in the developing foetus have many effects, and some of these involve the development of the brain.

Sex Determined

Females carry two identical sex chromosomes, XX, in their cells, while males have different sex chromosomes, XY. This means that all eggs carry X chromosomes, while 50 per cent of sperm carry the Y, so the sex of an embryo is determined only by the genetic contribution from the father. The Y chromosome has a unique gene that encodes for a chemical protein called testis-determining factor (TDF). Between the sixth and twelfth week of a human pregnancy, this factor affects the immature reproductive systems. The sex organs develop into testes, which link to the urinary system by sperm ducts, and a penis develops. Without the factor, sex organs become ovaries, fallopian tubes develop, and the system develops a womb, vagina and clitoris. Once the sex organs are in place, these produce the first sex hormones: more androgens such as testosterone in males, and more oestrogen in females.

As the brain develops, other sex-specific differences begin to make it respond differently to these hormones. The brains of males produce more receptors that bind to testosterone, while female brains end up with more receptors for oestrogen. In this way, during the second half of a pregnancy, the brain becomes either "masculinized" by the

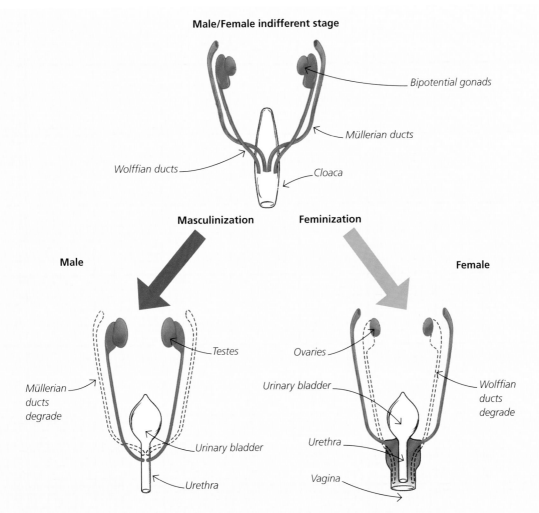

Male/Female indifferent stage

Bipotential gonads

Müllerian ducts

Wolffian ducts

Cloaca

Masculinization

Feminization

Male

Female

Testes

Ovaries

Müllerian ducts degrade

Urinary bladder

Wolffian ducts degrade

Urinary bladder

Urethra

Urethra

Vagina

The sex chromosomes in cells of the embryo determine the path of development from an undifferentiated state (middle). An embryo under the influence of testosterone forms male sex organs (left), while without the testosterone they become female (right).

effect of male hormones or "feminized" by the effects of female hormones. The exact significance of these physical differences between the brains of male and female foetuses, and their effects on later development and behaviour, are still a matter for debate and further research.

Sexual differences that begin with chromosomes continue through development as the bodies of males and females, both before and after birth, respond to the differences in sex hormones that now circulate through their bloodstream. Although testosterone and oestrogen are produced by both sexes, their relative amounts differ greatly. Testosterone is the basis for masculinization, while oestrogen controls feminization. This process also initiates chain reactions involving different kinds of hormones. Many of these are produced by the brain (or structures such as the pituitary gland that are associated with the brain), or interact with the brain to affect the way it develops.

Sex hormones have a profound effect on the development of the brain inside the womb. Testosterone and oestrogen are both present in all brains, but one or other dominates depending on the sex of the individual.

A Developing Brain

The fact that the sex hormones testosterone and oestrogen bind to receptors in the brain in the way they do suggests that they will influence behaviour. In the second half of pregnancy, this is reflected in the way they help to shape future sexual behaviour: they are involved in producing sexual identity and sexual attraction (see next section). But they may go much further in affecting other aspects of behaviour. Testosterone increases aggression and affects the way we interact with other people. Some scientists suggest that, in our prehistoric ancestors, this would have been important in helping to mould sex-specific roles in

the primitive social group, where aggressive males were needed for defence and hunting, while females devoted their time to rearing children. Today, these stereotypes are less applicable, but the physiological effects of the sex hormones persist.

These differences in behaviour began in the womb, when a balance of testosterone and oestrogen bound to receptors in the child's brain. An additional surge of testosterone occurs in the third month after birth. Studies suggest that testosterone has an effect on the way the higher parts of the brain develop. For instance, it may stimulate the amygdala, the so-called "fear and aggression" centre (see chapter 9), to develop. But it may inhibit certain forms of development in the prefrontal cortex, the region involved in "reading" the emotions of others (see chapter 10). This means that, even in the early stages of childhood development, the sex hormones are directly influencing deep parts of the brain involved in emotion and temperament.

How the Brain Affects Puberty

Testosterone and oestrogen are steroids that are produced by sex organs (testes and ovaries), but a separate class of hormones that are proteins are produced by the core of the brain to control sexual development. They are the so-called gonadotropic hormones. Their name means "hormones that nourish the gonads" and they target the sex organs, bringing girls and boys to sexual maturity so that they can produce children of their own. What happens at puberty is dependent on a complex interaction between the sex hormones coming from the sex organs and the gonadotropins coming from the brain. During this time, the hypothalamus in the brain's core produces a surge of gonadotropin-releasing hormone, GnRH, which travels to the front of the pituitary gland (the body's "master" hormone-releasing gland) just below the hypothalamus. This gland, in turn, releases the gonadotropin hormones necessary to activate the sex organs. As a result, these parts of the body produce their own surge of oestrogen or testosterone, and this hormonal flow triggers the onset of secondary sexual characteristics, such as body hair and the growth of external genitals.

Most importantly, the effect of gonadotropins is to bring the sex organs into maturity so they are capable of producing sex cells. In boys, the result is an uninterrupted production of sperm that will carry on through their life. In girls, it is the start of a more complex monthly cycle.

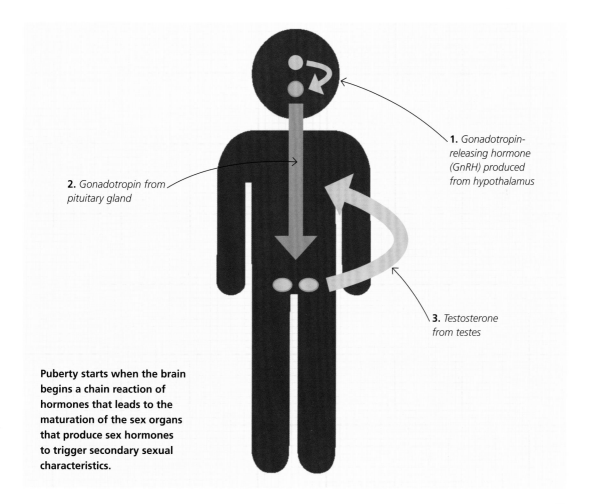

1. *Gonadotropin-releasing hormone (GnRH) produced from hypothalamus*

2. *Gonadotropin from pituitary gland*

3. *Testosterone from testes*

Puberty starts when the brain begins a chain reaction of hormones that leads to the maturation of the sex organs that produce sex hormones to trigger secondary sexual characteristics.

Investments of the Sexes

Many biologists think that the human menstrual cycle evolved as a way of reinforcing the bonds between potential mother and father. Virtually all other mammals go through a period of oestrous in their sexual cycles. This is where the female advertises when she has eggs that are ready for fertilization by changing her appearance or behaviour. Famously, many monkeys develop colourful bottoms to show when they are in oestrous, or "season". Prospective human fathers have no such flag, and it may be that, at least in our primitive ancestors, it would have paid for them to stay around for longer. That, in itself, is enough to mould the developing human brain, perhaps making men more receptive to long-term partnerships. But why is there a cycle in the first place?

Across the animal kingdom, females, in general, invest more in the next generations than males, although the exact level of male involvement varies widely. This is, perhaps, at the very core of sex-specific differences in behaviour. It costs a mother more physically to produce a child than it does a father. In all animals, eggs are bigger than sperm (in humans, eggs are the biggest cells, while sperm are among the smallest) because eggs are endowed with a greater quantity of cytoplasm, with the addition of some yolk. Sperm, in contrast, are little more than swimming sacs of genetic material. This is offset by the fact that males produce more sperm than females produce eggs, but it is still costlier in terms of energy to be female, especially if the fertilized egg is going to be nourished in her womb. In mammals, pregnancy zaps the mother of much of her food and oxygen as her placenta feeds the growing foetus. And, while her brain prepares her to produce milk, the overall toll on her body ends up being huge.

Animals that scatter eggs (such as most insects and fishes) can afford to reproduce continually. But human mothers must pace themselves. It takes time to rear a baby, and her womb is best suited to accommodate just one growing foetus at a time.

On the Great Barrier Reef, off the coast of Queensland, Australia, a male pink anemone-fish guards a clutch of eggs. He invests energy to protect the eggs, but once they hatch, the young are on their own.

The Brain, Sex and Gender

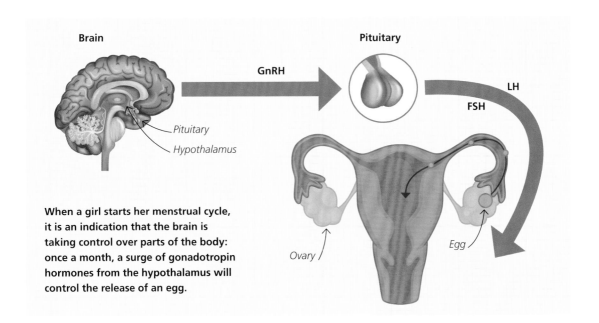

Brain

GnRH

Pituitary

LH

FSH

Pituitary

Hypothalamus

When a girl starts her menstrual cycle, it is an indication that the brain is taking control over parts of the body: once a month, a surge of gonadotropin hormones from the hypothalamus will control the release of an egg.

Ovary

Egg

Typically, just a single microscopic egg is released in a monthly cycle, and, if it is fertilized by sperm, we shall see that the embryo itself interferes with the natural cycle to stop any more eggs from being released for the duration of the pregnancy.

How the Brain Controls the Menstrual Cycle

A baby girl is born with ovaries that already contain all the eggs that she will ever release in her lifetime. But these eggs are immature, and they need the gonadotropin trigger from the brain to develop and release them. At the onset of puberty, the brain produces two main kinds of gonadotropins. They come in a monthly surge, one following the other. The first is called follicle-stimulating hormone (FSH); the second is luteinizing hormone (LH).
Both flow from the front of the pituitary gland in response to specific releasing-hormones (the GnRHs) coming from the brain's hypothalamus.

The first surge in FSH activates eggs inside an ovary. It makes them develop inside growing fluid-filled sacs called follicles. The second surge of LH then triggers one of the follicles to burst from the ovary, releasing an egg into the fallopian tube. Together, therefore, the brain's gonadotropins

have worked to cause ovulation. But while a follicle develops, it acts as its own hormone-producing gland: it adds to the level of oestrogen in the body. This rise in oestrogen feeds back to the hypothalamus to stop any further production of gonadotropin. At this stage, the body is preparing to get pregnant, so no more eggs are needed. The raised oestrogen also has another effect: it stimulates the wall of the womb to thicken so it can receive a fertilized egg during implantation. By the time the egg is moving down the Fallopian tube (where it might be fertilized), the ruptured remains of the follicle are adding yet another hormone to the mix: progesterone. This continues to inhibit the brain's gonadotropins and helps to maintain the thickened womb lining.

Everything stops after a few days if the egg isn't fertilized. Levels of progesterone fall as the follicle remains disintegrate, which makes the blood-filled womb lining break away: causing menstruation. This breaks the preparation for pregnancy. The brain is no longer inhibited, so the gonadotropin surge can happen again. Roughly a month after the last egg was activated, a new one stirs in the opposite ovary.

The embryo is just a ball of cells when it lands on the thickened wall of the uterus. It releases a hormone that prevents the mother from menstruating.

How Pregnancy Breaks the Brain's Menstrual Control

If an egg is fertilized, the prospect of a pregnancy means that the hormonal interplay between the brain and the ovaries must be interrupted. It is the embryo itself that causes this. After implantation in the thickened wall of the womb, the embryo starts to produce a hormone of its own, called HCG (human chorionic gonadotropin), from its surface layer of cells that is now embedded within the tissues of the wall. HCG preserves the ruptured follicle, meaning that it keeps producing progesterone. This means that the brain's gonadotropins continue being inhibited and the ovaries are not reactivated to release any more eggs. This happens as long as the pregnancy progresses. The monthly cycle of gonadotropins will only start up again after the baby is born and the brain is no longer inhibited.

Hormonal Balances

The hormonal sex-specific cocktails that are mixed before birth will affect us throughout our lives. As we have seen, many of these hormones are produced by the brain itself. Others are produced by the sex organs. A complex interplay between all of them keeps their levels in check, stopping them from rising too high or dropping too low. Regulation happens through a process of negative feedback: any deviation away from an optimum level, or norm, triggers a response that corrects it. Negative feedback is central to keeping conditions right inside the body (see chapter 6). The brain and sex organs have a two-way system of communication

The Brain, Sex and Gender

that will continue through life. When blood containing testosterone and oestrogen released by the sex organs reaches the pituitary gland, the hormones depress the gonadotropins. Then, when these drop, so does the testosterone and oestrogen. In this way, sex hormones from the brain and gonads help to regulate each other, so the ultimate effect is that their levels never rise too high or drop too low. At puberty, this balance between the hormones is effectively reset so the body can sustain the surge in testosterone or oestrogen to bring about sexual maturity. And in girls, the cyclic interplay between the two starts to take the form of the monthly menstrual cycle.

Hormones and Behaviour

Hormones affect our behaviour because they affect our brains. And the sex hormones that change our sex organs and physical characteristics can also impact on the brain because all these sites carry the receptors that are specific to these hormones. The hormones must bind to these receptors to exert their influence. As well as the steroid sex hormones, oestrogen and testosterone, from the ovaries and testes, there are the other protein-based hormones produced by the brain that influence females and males in different ways. Some of these, such as the gonadotropins, are made in the pituitary glands, but are triggered by releasing factors from the brain's hypothalamus. Others are produced by the hypothalamus itself. One of these is oxytocin.

Oxytocin plays a particular role in sex-specific behaviour. As we have seen (chapter 6), it has been called the "love hormone" because it encourages social bonding in a number of contexts. Its most notable effects at the very start of life are far more physical. It stimulates muscles of the womb to contract during labour, and oxytocin from both mother's and baby's brains help with this. After birth, oxytocin will then go on to stimulate the mother's breasts to release milk for feeding the baby. Mothers experience a surge of the hormone when the baby sucks on a nipple, but the fact that oxytocin also binds to higher parts of the brain

means that it has a broader influence on maternal behaviour. Oxytocin seems to depress the effects of stress hormones such as cortisol, and it clearly has an influence on parts of the brain involved in controlling moods and emotion (see chapter 9). It depresses the brain's amygdalae: the paired almond-shaped structures that are centres of aggression and fear. A mother receives a rise in oxytocin when she plays with her child, making her feel happier and calmer. And the oxytocin high is reciprocated: children experience a rush in oxytocin when comforted by their mothers. Even beyond the mother–child relationship, oxytocin exerts its effects: it rises during any friendly social interactions, stirring feelings of trust and generosity. Conversely, women who are emotionally stressed, such as through abuse, have low levels of oxytocin in their bloodstream.

Changes in hormone levels can cause us to experience an emotional roller coaster.

Unlike the lingering effects of stress hormones, pleasure-enhancing surges in oxytocin do not last long, but in a friendly social setting, levels of oxytocin might be continually boosted. Oxytocin has a stronger impact on females because in males it is balanced with the higher levels of testosterone. But men can also experience oxytocin highs, especially during fatherhood, when they play with or comfort their children. At times like this, their usual testosterone levels are depressed. Studies of hunter-gatherer societies have shown that, while oxytocin levels in men are low during the thrill of a hunting expedition, they rise when the men return home, helping them to reconnect with their families.

Another hormone, called prolactin, produced by the front pituitary, has a similar effect to oxytocin, but in a somewhat specific way. Prolactin is produced by the front of the pituitary gland and stimulates the mother's production of milk before oxytocin can help with its flow. But it is also implicated in encouraging "nesting" behaviour. Both parents can experience a surge in prolactin, and they not only become more protective of their child, but are also inspired, for instance, to prepare the house so it is ready for the new arrival.

In some cases, the effects of hormones can be particularly sex-specific. Vasopressin is a hormone produced by the hypothalamus (and, like oxytocin, released from the rear pituitary) that affects everybody by encouraging water retention in the kidneys. It therefore flows more when we are dehydrated, such as after exercise (see chapter 6). But it also influences defensive behaviour. It makes mothers more protective of their offspring (even in a potentially aggressive way), but makes them better able to read facial expressions, so recognize and trust friendly strangers. By contrast, in men it appears to be more likely to make them see strangers as hostile and threatening.

Studies suggest that the production of the hormones prolactin and vasopressin can make mothers more protective of their children.

The Brain, Sex and Gender

The urge to "nest" – to create a comfortable environment for a newborn child – has been linked to a rise in the hormone prolactin.

Sex-Specific Brains?

This section shows how females and males are physically and behaviourally different because of the genes and chromosomes that influence development. And this development includes the ways they produce and respond to the chemical signals called hormones. But are the brains of women and men structurally different?

There are no aspects of brain structure that are diagnostic between the genders. In other words, it is impossible to be sure of the sex of an individual just by looking at the size and shape of the brain and its parts. And there is no structure that is uniquely found in one gender or the other. Rather, there are differences in averages, with considerable ranges of overlap. For instance, men seem to have on average bigger amygdalae than women: the so-called "centres of aggression and fear". One study found that women have stronger connections between the hemispheres than men, whereas in men the connections within each hemisphere seem to predominate. Some have taken this as evidence that the different sexes think in different ways and that, for instance women are better able to combine analysis with intuition (see chapter 7 for more on "right and left brains"). But many studies that have documented supposed differences in brain structure have been inconclusive or even contradictory. Even the presence of hormones themselves is effectively gender-neutral. Both sexes have quantities of oestrogen and testosterone, but the balance between the two is important. It is clear that sex-specific functions of the brain depend on the balance of chemical mixtures. In other words, whether we are female or male might at its root be more about chemistry than anatomy.

The Brain and Sexual Behaviour

As the human brain matures, both in the womb and after birth, it is influenced by genes and hormones in ways that have a profound influence on sexual behaviour.

The brain is the sexiest organ of the body. Not only does it help shape our maleness and femaleness, but it also controls our sexual drive and determines the kind of person that we are sexually attracted to. It makes us fall in love, and fall out of love. It makes us infatuated or helps to bond us into stable relationships. And sometimes it makes us jealous. During the sexual act itself, the brain is actively involved in giving us a feeling of euphoria. And, as we have seen in the previous section, it even prepares us along the road to raising a family.

Our brains have evolved to make sex a pleasurable experience, so that we are inclined to do it again and again. In fact, sex and reproduction are such fundamental characteristics of life that they are controlled by the same parts of the brain that are involved in other vital regulations, such as the brainstem and core of the cerebral hemispheres.

Falling in Love

Pleasurable feelings of being in love involve parts of the brain that are at the root of other rewarding sensations, such as eating good food. This reward system is part of a region of the brain called the limbic system, and it involves the release of the neurotransmitter dopamine (see chapter 9). While it lasts, this overwhelms more rational parts of the brain, such as the prefrontal cortex. This means that, just like other cravings, we are being controlled by a deeply-seated automatic part of the brain that is driving us to satisfy the craving. The limbic system of our brain motivates us to pursue the object of our infatuation in order to satisfy its pleasure centres. At the same time, levels of the stress hormone cortisol rise. Cortisol reduces the levels of testosterone in love-struck males, but in love-struck females, testosterone levels go up.

All animals, including giant tortoises and humans, are adapted to pass on their genes by reproduction. Reproduction is at the core of life itself, so our bodies and brains have evolved to increase the chances of this happening – even with a hard shell in the way!

Choosing the Right Mate

Mate choice is an important business. Among animals, a wide variety of mating strategies have evolved to ensure that they have a good chance of passing on their genes. And in the general game of picking mates, females are usually choosier than males. The reason comes down to the different investments that mother and father make in raising a family. Since eggs are usually costlier to make than sperm, it makes sense for females to take more care over their choice of mates. They want to ensure that the males that fertilize the eggs provide them with the best genes for helping them to grow into healthy babies. The animal kingdom is filled with examples of how this female choice is manifested. Males, in general, do their best to impress the females. Birds may do it with colourful plumage or loud songs. Even though these same characteristics are likely to draw the attention of dangerous predators, males still flaunt what they have: they want to persuade the females that they have good genes and are strong enough to take the risk. Male bower birds attract females by building elaborate display areas called bowers. The males meticulously care for their bowers, ensuring that everything is in the right place. Females visit bowers in turn and give them a good inspection while the males display for them. Those males that produce the most impressive bowers and displays are the ones that get to mate.

In humans, secondary sexual characteristics may serve the same purpose: as signals that attract the opposite sex. But are female humans just as choosy? Some research suggests they are: that women are more likely to be drawn to men who are more likely to make a success of raising and caring for the family. But there are hidden complications, including some evidence that body odour may be a cue. MHC (major histocompatibility complex) is a set of proteins that is important in the immune system. Varieties of MHC are expressed by different varieties of genes, and studies suggest that women may be attracted to men who are genetically different from them in this respect. They do it by subconsciously choosing an MHC-specific body odour. This might be a strategy that helps humans to avoid inbreeding with individuals who are genetically related. Mating between the most genetically unalike individuals, with different MHC proteins, is more likely to produce genetically healthy babies.

A male bower bird carefully tends to its bower as it tries to woo a mate.

Over time, as the bonding develops, the role of the limbic system wanes and the prefrontal cortex gains greater influence. Cortisol levels drop. When this happens, we are in a better position to make reasoned decisions on the appropriate path to take with the relationship. The process of courtship can be seen as a strategy for maximizing reproductive success. At first, the limbic system drives us to test a partnership before the prefrontal cortex comes into play to help make the choice. Courtship is a process that happens, on a more fundamental level, in all kinds of animals. It serves to help the male or female decide whether the partnership that lies ahead will be the best one for rearing a family. Once the pairing has formed, hormones such as oxytocin will help to reinforce the bond.

Sexual Arousal

During sexual intercourse, the brain's pleasure centres are accompanied by impulses that are firing through the body's automatic control system. This is the autonomic nervous system (see chapter 6). It is responsible for regulating vital functions, such as heartbeat, breathing and digestion, and its control centre is located right at the base of the brainstem. The autonomic nervous system is entirely involuntary, meaning that we have no conscious control over parts of our sexual stimulation. But sexual stimulation must begin with conscious actions and sensations. Like most routine sensory information, impulses generated from touch, vision and so on travel to the main sensory hub of the brain: the thalamus at the core of the cerebral hemispheres. From there, they affect the dopamine-producing parts of the limbic system. This is what gives us the pleasurable feeling. The amygdalae, however, are inhibited: these areas of the brain ordinarily suppress our sexual feelings. Motor impulses then travel to the genitals to make them sexually aroused, and the automatic nervous system triggers orgasm. At the same time, there is a rush of oxytocin, the "love hormone", from the brains of both partners. The oxytocin stimulates muscles in the reproductive tracts to contract: in the man this helps to squeeze sperm along the direction of the ejaculation. In the woman, it stimulates the Fallopian tubes to bring the egg down closer to the womb, ready to be fertilized.

Gender Identity

As we have seen, the sex hormones oestrogen and testosterone play an important part in making us "masculinized" or "feminized", not only in our outward appearance but also in how we feel. While our biological sex relates to the kinds of bodies we have, our gender identity concerns whether or not we think of ourselves as "girl" or "boy". Traditionally, toddlers were thought to acquire their sense of gender identity by the age of three years, learning that they were either a "boy" or a "girl" as they developed and started to make sense of their place in the world. However, recent research suggests that the process by which we acquire a gender identity begins much earlier than this, while the baby is still in the womb during the second half of pregnancy.

In some cases, a medical condition can result in a mismatch between an individual's biological sex (as determined by the XX and XY chromosomes), their hormonal sex (determined by the masculinizing or feminizing effects of testosterone of oestrogen) and their gender identity (whether the individual "feels" male or female). For instance, sufferers of androgen insensitivity syndrome are born with the male complement of chromosomes, XY, and produce normal amounts of masculinizing testosterone, but their developing bodies fail to respond to the testosterone. As a result, their sex organs and brain are feminized. They will grow up looking outwardly like women, and will often develop a strong sense of female gender identity, even though they have XY chromosomes. Inside, they lack a uterus and do not menstruate, which means that they are infertile.

In other cases, a person is said to be suffering from "gender dysphoria" when their sense of gender identity does not match up with their biological sex. Such people may choose to live with the "opposite" gender identity: for instance, living as a man despite having a female body. They may also decide to receive gender-reassignment treatment, including sex hormones, to bring their bodies closer into alignment with their gender identity.

Evidence that gender identity may begin to develop as early as inside the womb can come from specific case studies. One sad case study that supports this idea involves a boy who lost his penis aged just eight months when a circumcision went badly wrong. Together, his doctors and his family decided that it would be best to also remove his testicles and to raise him as a girl. The psychologist involved in the decision was confident that gender identity was determined solely by upbringing and environment, and that the child would identify as a girl, but this did not happen. Rather, the child grew up unhappy with the transition, and with strong feelings of being male. Eventually the change was reversed. While the scientific debate over the development of gender identity continues, this case provides evidence against the idea that gender identity can be changed at will after birth.

Sexuality

Humans can develop a wide variety of sexual orientations regarding those to whom they are sexually attracted. The people we have been considering so far have generally been heterosexual, meaning that they are sexually attracted to the opposite sex. But some people are homosexual, meaning that they are attracted to members of the same sex, while others are bisexual, meaning that they are attracted to members of both sexes. As with gender identity, there is evidence that our sexuality may start to develop in the womb. Genetic studies, including the finding that homosexuality is often shared by identical twins, suggest that sexuality may be at least influenced by different combinations of genes

Gender dysphoria occurs when the sense of gender identity that has formed in a person's brain does not match up with their physical biological sex.

on our chromosomes. Such genetic features affect brain development, but much about the development of sexuality remains poorly understood, including the role played by the environment in which we grow up. The hypothalamus and amygdala may be involved in shaping our sexual desires, and certain structural differences in parts of the hypothalamus have been linked to sexuality. However, different studies have found conflicting results, and research is ongoing to make sense of their findings. The ways in which hormones and brain structure come together to shape gender identity and sexuality are complex, and the interaction between the brain and the environment also needs to be considered to form a full picture of the ways in which complex characteristics such as these are formed. However, it is clear that it is in the workings of our brains that such characteristics are ultimately expressed.

The diversity of human sexuality includes heterosexual, homosexual and bisexual attraction. Recent studies suggest that there may be differences in brain development between heterosexual and homosexual people, while twin studies suggest that the influence of environment is also important. How these interrelate is the subject of ongoing research. There remain many competing theories regarding the development of sexuality in the brain, but the growing evidence that sexuality is not a choice has helped homosexual people to win legal rights around the world.

The Brain, Sex and Gender

Chapter 12
THE BRAIN THROUGH LIFE

The Brain Inside the Womb

It takes 40 weeks to grow a human brain, at least to a point where its owner can survive independently of their mother's body. During that time, the brain develops all the parts needed to keep it working through the rest of its life.

Inside the womb, a human embryo grows gradually, moment by moment, not in fits and starts. Like every other organ of the body, the brain also grows gradually. It develops out of a group of cells that shuffle into position at a time when the embryo is far too small to see with the naked eye. The cells divide and the brain gets bigger, along with the rest of the embryo. Therefore, there is no fixed time during the embryo's development that we can say that the brain appears. Indeed, in one sense, it has been there all along, at least in potential, as the thousands of genes found inside the fertilized egg. These will collaborate to bring the necessary cells all together and keep them working along the right lines to build the human brain.

Development of the embryonic brain follows a similar path to the evolution of the brain in the prehistoric past. At first, the embryo concentrates on putting down the ground roots of the brainstem that will be so important in controlling vital functions, such as the first heartbeats. Then it will work to develop the parts of the brain that give sensitivity and feeling.

By the time a baby is born, its brain is already recognizably human. The brain is still only 25 per cent of its final adult size, and will continue to grow rapidly.

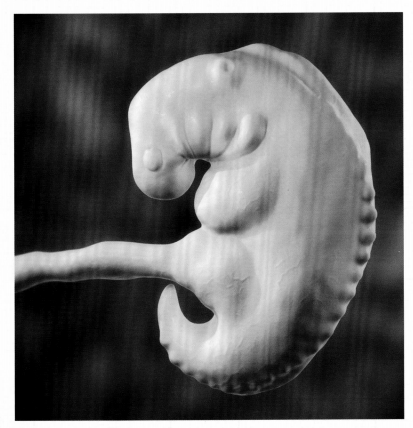

A human embryo just after week four is only 2 mm (¹/₁₂ in) long and shows the knobbly components of its spine running along its back. Here, the spinal cord, which formed first inside a furrow, is closed over and protected inside the vertebrae of the developing backbone.

Sowing the Seeds of a Human Brain

During the first three weeks of its development, the human embryo is little more than a microscopic ball of cells embedded in the blood vessel-filled lining of its mother's womb. After a few days of being solid, it becomes hollow and its outer layer helps it attach to the womb. Deep inside the hollow ball, sandwiched between two tiny fluid-filled bags (one of which becomes the amniotic sac that will protect the foetus in the womb), there is a little disc of cells. This specific part will develop into the baby. After the embryonic body thickens and grows the rudiments of its gut cavity, the first signs of a nervous system appear. Running in a line through the top of the embryo is a strip of cells that grows inwards and folds around to form a tube. This process is called neurulation and it

starts just before week three and continues into week four. This so-called neural tube will become part of the baby's central nervous system: its spinal cord. The front of this tube widens and bends forwards into the head as it grows into a brain.

By the start of the fifth week, the main regions of the brain (called vesicles) have differentiated. The brainstem runs from the base of the brain's swelling, where it joins the spinal cord, to the region of the forward-bend. Specifically, the part before the bend is the early hindbrain, while the region of the bend itself is the midbrain. Both hind- and midbrain will be concerned with controlling the vital organs: see chapter 5. In front of the fold, the brain swells to form the first forebrain, which will later come to be dominated by the cerebral hemispheres.

The Early Brain

By week six, the embryo looks like a fat tadpole, but is no bigger than a lentil. Inside, its brain looks like a twisted worm with a swollen end, but it is already firing impulses. The swollen end will grow into the cerebral hemispheres. Half of the length of the entire brain up from the base is made up of the primitive hindbrain, curved slightly inwards from the back, while the bend of the midbrain arches round even more. The brain's cranial nerves have also grown. The sense organs, muscles and glands that the nerves connect to are not properly formed yet, but the embryo's facial features – its mouth and nose – are starting to take shape. One pair of cranial nerves, the optic nerves, are clearly joined to the embryo's first eyespots, or optical vesicles. These will grow into the eyes, although it will be a long time before the baby can see. Indeed, most of the embryo's senses are not yet developed. A sense of touch is the first to appear, emerging around week eight, but, in the warm, dark confines of the womb, senses are largely unnecessary. And although the muscles and the skeleton the muscles are connected to are growing, the embryo won't start moving until about week eight. Even then, the embryo is still so small that the mother won't yet feel it.

Already the brain is involved in keeping control of what the body is doing. The heart was the first organ to be formed in the embryo, and started beating at the beginning of the third week. Unlike other muscles in the body, heart cells contract spontaneously: they don't need nerves to stimulate them. But as the brain develops its autonomic systems for regulating organs, the heart is brought under its control. The medulla oblongata formed in the base of the primitive brainstem connects to the young heart with nerves that will keep it beating within reasonable limits. It starts out with a speed that approximately matches that of the mother. It then gradually gets faster until it more than doubles this speed around week nine, before declining slightly again after that.

By week seven, the foundation of the embryo's body parts has been laid down, but the brain remains primitive, with a long worm-like brainstem and bulbous forebrain.

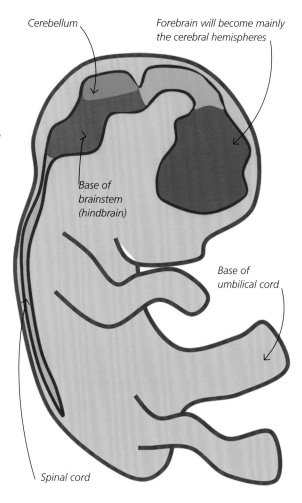

Cerebellum

Forebrain will become mainly the cerebral hemispheres

Base of brainstem (hindbrain)

Base of umbilical cord

Spinal cord

Although the embryo's nervous system is controlling the embryo's body, this is happening at an automatic subconscious level. In order for it to be consciously aware, the brain's "higher" parts need to develop. As we have seen, these require the brain's cerebral hemispheres, especially the surface layers called the cerebral cortex. These layers have not yet developed.

Origins of the "Thinking" Brain

The cerebral hemispheres dominate the adult brain, making up the largest share of the total brain volume. As we have seen, their surfaces are folded and grooved to pack in as much surface grey matter as possible (see chapter 7). It is this grey matter in the cerebral cortex that is so critical for conscious, rational thinking. Our cerebral hemispheres form when the swollen front end of the embryonic brain – its forebrain – divides down the middle along a furrow called the longitudinal fissure. This starts to happen when the embryo is five weeks old, and continues up to week ten. By then the baby, now called a foetus, is about the size of a large olive and is more recognizably human in shape. It has translucent skin and tiny bending limbs, with fingers and toes that are starting to grow nails. It has a distinct human-like face, complete with nose, eyelids and earlobes.

At first, the primitive cerebral hemispheres are smooth and largely undifferentiated. But from week ten they embark on a process of reshaping that will be critical to forming the thinking human brain. During this time the stem cells that will become neurons divide and proliferate. They migrate to new positions as the cerebral hemispheres are reorganized to make their new circuits. By week 12, the first outward signs of this reshaping are seen when the side of each hemisphere develops a new groove, called the Sylvian fissure (named after

Franciscus Sylvius, a seventeenth century Dutch physician who is credited with its recognition). Each fissure marks the fashioning of the cerebral hemisphere's first lobes, as the frontal lobe separates from the temporal lobe on the side. The frontal lobes will become involved with complex thought processes, such as reasoning, whereas the temporal lobes will be areas for processing sound and language. Around the same time, the band of white matter, the corpus callosum, grows between the two hemispheres to help them communicate with each other. Already the brain is laying the ground plan for complex conscious thought, although its cerebral cortex is still too underdeveloped to perform these tasks. The foetus is still responding by automatic reflex actions. It cannot direct any of its actions in a purposeful way, but the surface of the cerebral hemispheres is already starting to develop folds: the gyri that will accommodate the necessary surface grey matter to do this. This is reflected in better coordination between senses and movement. By week 15, the foetus shows movement of facial muscles and may even suck its thumb. And even though the eyelids are shut, it may respond to the beam of a flashlight directed onto the mother's abdomen. Shortly afterwards, the mother may feel the foetus moving and kicking, and the foetus will hear the sound of her voice.

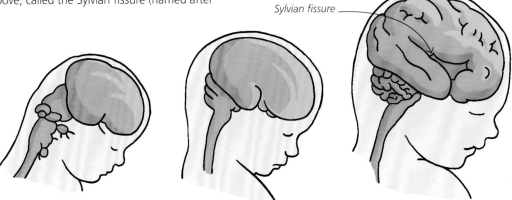

Sylvian fissure

From around week 12, it takes more than seven weeks for the main lobes of the cerebral hemispheres to form. During this time, their cerebral cortex will show signs of forming their first folds and grooves to accommodate more surface grey matter.

Wiring Up the Thinking Brain

The key to complex brain functions is the complicated connections between neurons. As we have seen, neighbouring neurons communicate via their spindly nerve fibres at junction points called synapses (see chapter 2). This happens when the tip of one fibre releases a chemical signal called a neurotransmitter that binds to the next neuron. Some neurotransmitters excite activity, other inhibit them. But it is the overall pattern of synaptic connections and balance of neurotransmitters that determines what the brain can do. In the cerebral cortex, this pattern is very complicated indeed.

Lots of new synapses appear in the foetus's brain by week 17. At the same time, the cerebral hemispheres develop more folds (gyri) and grooves (sulci): it becomes more like the adult human brain in outward appearance. By week 23, now about the size of a large mango, the foetus is capable of surviving outside the womb with medical support, and it responds to irritating stimuli. Some of the brain's nerve fibres develop their fatty layers of insulation (the myelin sheath: see chapter 2) that will help to speed up the movement of nerve impulses. Overall, there are signs that the cerebral cortex is starting to exert its control of conscious, voluntary action. But parts of the brain are still underdeveloped. The biggest peak in growth of synaptic connections doesn't happen until week 28. And the hypothalamus will not be capable of self-regulating body temperature until week 32. This is when the brain starts to control breathing movements: the foetus inhales and exhales the womb's amniotic fluid, which helps to develop the lungs.

The Brain Through Life

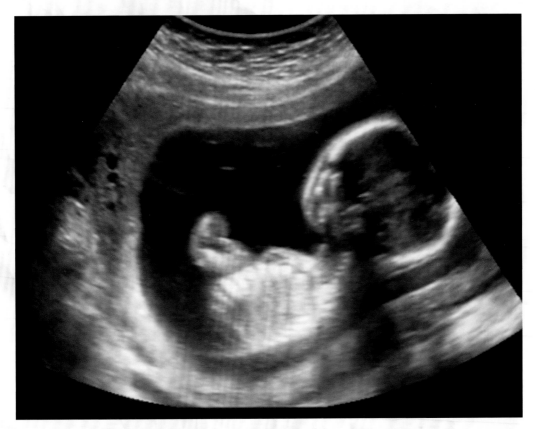

This ultrasound scan of a foetus during the sixteenth week of the pregnancy clearly shows the skull with the developing brain inside.

Influences of Maternal Blood

As the embryo grows inside the mother's womb, it is supported by an organ called the placenta. The placenta, in fact, is a kind of hybrid structure, because part of it is formed from the embryo's tissues and part comes from the mother. It serves as the interface between the bloodstreams of the two and controls which substances can pass from one to the other. All the embryo's nourishment and oxygen come from the mother across the placenta. Waste products, such as carbon dioxide, move back the other way. For the embryo, the placenta is its life-support system.

The boundary between the two bloodstreams allows the free exchange of chemical molecules, but not cells. This means that hormones from both mother and baby mix together, but also more damaging things like pollutants and drugs. Many of these can interfere with the way the embryo's brain develops, and some can lead to irreparable damage. Alcohol, for instance, can disrupt the way neurons migrate when they are forming the critical layers of the cerebral cortex. This can lead to foetal alcohol syndrome, where the baby is born with a brain that is under-sized, or left with mental problems that lead to difficulties with learning and behaviour. Similarly, smoking can increase the risk that children will develop aggressive behaviour or speech problems, while prescribed drugs such as antidepressants can even cause physical defects such as spina bifida. These effects show just how vulnerable the developing brain can be.

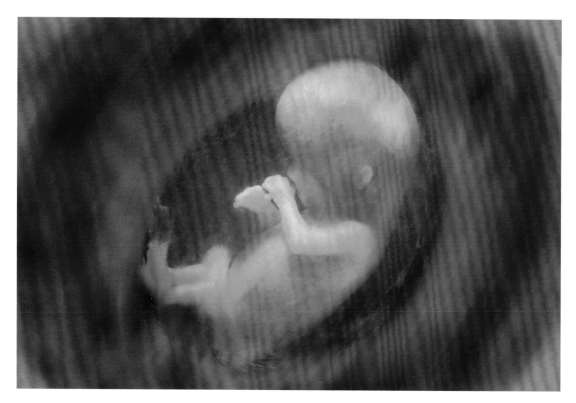

A foetus starts to suck its thumb from the early stages of pregnancy. As it develops, it becomes better at doing this, learning to open its mouth in anticipation and to move the thumb accurately to the right place. The hand the foetus sucks is normally the one it will favour throughout its life.

Becoming Consciously Aware

By 30 weeks, the cerebral cortex has formed from a layer of neurons called the cortical plate. Meanwhile, deeper inside each hemisphere the brain has formed the pockets of grey matter that will be important for passing information to and from the cerebral cortex. These make up the core of the forebrain that will help with initial processing of information, storing memories and provoking feelings of pleasure and discomfort. Central to each hemisphere is the thalamus, which acts as the hub for receiving sensory information (see chapter 7). Around this is the limbic system, which is involved in feelings and memory (see chapters 8 and 9). But these processes can be automatic and happen independently of conscious influence. For

us to have conscious control of our feelings, and, indeed, for us to be consciously aware of sensations, they need to connect to the cerebral cortex. In the foetus, these connections grow between weeks 12 and 16, but that is before the cerebral cortex is properly formed. And it is notoriously difficult, probably impossible, to identify a time when the foetus is consciously aware. Some regard the onset of the sleep–wake cycle as an important clue. During late stages of pregnancy, the foetus sleeps for about 95 per cent of the time, so wakeful periods are characterized as short bouts of activity, but many doubt that even these can be equated to genuine conscious wakefulness of the kind experienced after birth. When a foetus becomes self-aware remains a debated issue.

Can a Foetus Feel Pain?

Pain is a very specific reaction that involves a particular circuitry of neurons. It is a warning system that things are not right: part of the body is being harmed or damaged, and a quick response is called for to remove the source of harm. Pain involves activating warning centres in the deep parts of the cerebral hemispheres (see chapter 9), and processing in the cerebral cortex allows us to experience it. In order for a foetus to feel pain, all of these components must be working and connected.

The deeper parts of the brain develop before the cerebral cortex, when the brain's basic circuitry is being formed. And these parts, notably the thalamus, are involved in receiving sensory information. Around the body, different kinds of sensors also form, including pain receptors called nociceptors. They fire nerve impulses to the thalamus, and these then move on to the warning centre called the cingulate gyrus. That is enough to generate a reflex response (so a foetus will automatically twitch when stimulated by a needle), but without connections to a functioning cerebral cortex, the sensation of pain is not consciously perceived. Before the development of the cerebral cortex, any sensory experiences are more like our knee-jerk reflex. Studies of blood flow and EEG measurements of electrical activity in the cerebral cortex of premature babies show that pain stimuli are sensed first around week 25, corresponding with the time that final connections are being made that make the cerebral cortex properly "wired up".

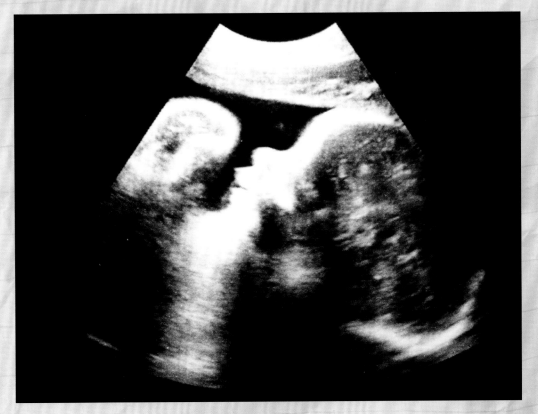

This ultrasound scan shows a developing foetus at about seven months. By this time, the foetus can sense pain as the final parts of the cerebral cortex are being connected.

The Childhood Brain

From birth to puberty, the human brain is constantly developing physically and mentally. It is accumulating experiences that will shape adult behaviour, and preparing the body for a life independent of parental care.

For all the development that has happened inside the womb, when a baby is born, its brain is still incomplete: it is only a quarter of its future adult weight. Compared with other animals, that makes the newborn human brain premature. But it has to be. If the newborn brain were any bigger, the baby's head would not be able to pass through the mother's pelvis during birth. A newborn brain has a phenomenal amount of growing to do to catch up, something that happens with remarkable speed after birth. Just within the first three years of a baby's life, its brain grows to 80 or 90 per cent of its adult size. Growth then slows as it reaches adult size around adolescence.

The wealth of new sensations experienced during childhood create many more connections within the developing brain, increasing its mass considerably. By age three, it has grown to 80 per cent of its adult size.

But, perhaps more remarkably, a baby is born with about the same number of neurons in its brain as in an adult. The increase in brain size is due to the proliferation of glial cells, and also to the process by which the neurons grow more fibres to make more synaptic connections. This wiring, and rewiring, continues until after adolescence, but it is especially important during childhood, when new sensations and experiences are helping to mould learning and behaviour.

The Brain's Role in Birth

Life inside the womb is calm: the foetus is cushioned inside its amniotic sac and generally buffered from outside influences. But as the pregnancy approaches its fortieth week, that starts to change. As the baby grows bigger, it demands more nourishment until a point is reached that the mother and her placenta can no longer satisfy it. The growing baby, subconsciously, begins to sense this: its blood sugar level starts to drop and this is detected by the brain's hypothalamus. This is a stressful situation, and the hypothalamus triggers a chain reaction via the pituitary gland that leads to the foetus's body producing cortisol, the stress hormone (see chapter 9). One of things cortisol does is to stop the placenta from producing progesterone. And that makes the womb more sensitive to another hormone: oxytocin.

As we saw in chapter 11, oxytocin is the "love" hormone. Produced by the hypothalamus and released from the pituitary gland by both the mother and baby, it strengthens the bond between mother and child. But it also stimulates contraction of smooth muscles of organs such as those of the womb. Contractions push the baby's head against the opening to the womb, the cervix, which, in turn, triggers the brain to release even more oxytocin. By the time this happens, birth is inevitable.

Birth, therefore, is triggered because of a subtle communication between the brains of mother and child. The delicate balance is necessary for labour to progress as quickly as possible, and babies with brain disorders that affect the hypothalamus experience longer and more difficult births. In fact, a number of brain conditions that are at least partly determined by genes, and so are present even before the symptoms later manifest, such as schizophrenia and autism, may be associated with longer and more difficult births.

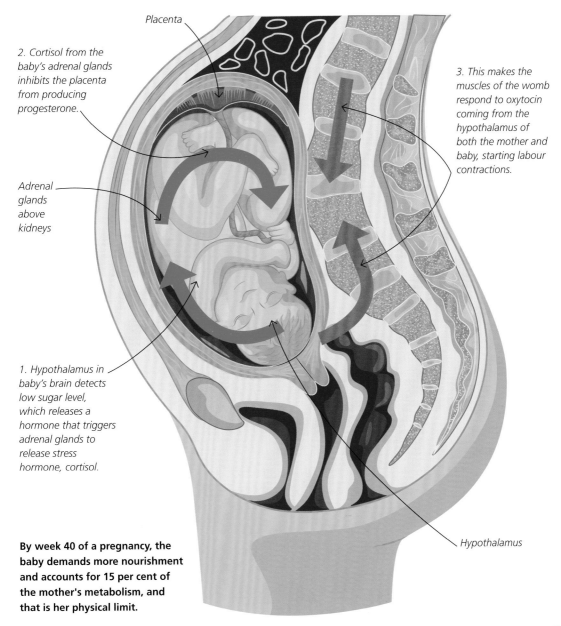

Placenta

2. Cortisol from the baby's adrenal glands inhibits the placenta from producing progesterone.

3. This makes the muscles of the womb respond to oxytocin coming from the hypothalamus of both the mother and baby, starting labour contractions.

Adrenal glands above kidneys

1. Hypothalamus in baby's brain detects low sugar level, which releases a hormone that triggers adrenal glands to release stress hormone, cortisol.

Hypothalamus

By week 40 of a pregnancy, the baby demands more nourishment and accounts for 15 per cent of the mother's metabolism, and that is her physical limit.

The Childhood Brain

The development of the hippocampus during the first three months of a child's life help it to recognize the sights, sounds and smells of its mother and father.

The Newborn Brain

A newborn baby is completely dependent on its parents. It sleeps in periods of up to a few hours at a time and must be fed regularly because its stomach is too small to accommodate bigger meals. Its behaviour is still dominated by the kinds of reflex responses that it had when it was still inside the womb. A rooting and sucking reflex ensures that the baby will first move towards a teat and then automatically suck to get milk. Its grasp reflex means it will grip objects with its hands. And although the senses, including vision and hearing, are working, the eyes cannot properly focus until around two or three months.

These behaviours are all innate: they are "hard-wired" in the baby's nervous system and are genetically determined (see chapter 8). For a time, at least, there are insufficient connections between the neurons of the brain for the baby to imprint memories that will help it learn much from its sensory experiences. As we have seen, memories are built as "maps" of neuron connections that arise as new synapses are made between their

fibres. However, there is some evidence that babies can remember smells from the womb. And shortly after birth, a baby might learn to be soothed by a particular piece of music. All this suggests that the first rudimentary memory "maps" might be made very early indeed. Within the first three months, the brain's hippocampus grows significantly. This is the region of the brain inside the cerebral hemispheres that is used for processing short-term memory into long-term memory.

New Connections: Blooming and Pruning

Automatic control systems continue to flourish as the child grows older. The cerebellum, the part of the brain involved in coordinating complex movements, triples in size in the first year to help with crawling and other motor skills. A big cerebellum will also be needed when the child starts walking upright. But, as we have seen, it is the cerebral cortex, the surface layer of grey matter on the cerebral hemispheres, that is so important in more conscious mental and cognitive development.

The Brain Through Life

For the first three years of life, the child's cerebral cortex develops more synaptic connections between its neurons than it will ever have in a lifetime. More fibres make the brain double in weight in the first year alone. While the number of neuron cell bodies stays about the same, the fibres that link them proliferate until, if we could see them down a microscope, they appear as a hugely complex, seemingly tangled, mass. The cerebral cortex is blooming. But these fibres and their connections are not at all random. They are made in precise ways in response to precise stimuli and experiences.

Inherited genes have determined the ground plan of the neurons from before birth, and therefore set limits on the ways in which they might connect. However, from the point at which a child starts to interact with its surroundings, memory and learning take over part of the control. Synapses are said to be plastic, meaning that they can be moulded by what the brain is doing. And as some new synapses are used more than others, these get strengthened. This means that the cerebral cortex is shaped through experience, and the more enriching the environment, the richer these connections and cognitive abilities will be. Research has shown that neglected children end up with shrunken cerebral cortexes and, overall, smaller brains.

By the age of three, a toddler's cerebral cortex has twice as many synapses as that of an adult. The reason for this is that any connections that are regarded as surplus to requirements are now trimmed away. This pruning happens throughout the rest of childhood and continues through to adolescence. It targets the synapses that are weakest, while preserving those that have been reinforced. This can affect the long-term wiring of the brain in many ways. For instance, by the end of the first year, repeated exposure to a particular language will help to "programme" this into the child's brain as its first language.

The number of neurons in the cerebral cortex does not change with age. They can be seen here as the dark spots of their cell bodies. But their fibres and synapses (connections) change considerably. These peak in number at around three years, corresponding to a massive potential for learning. Afterwards, the connections that are not used are pruned away.

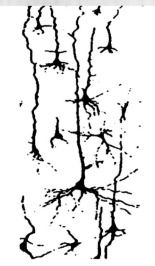

At birth　　　　**Six years old**　　　　**Fourteen years old**

The Childhood Brain

177

Using the Abundant Synapses

Around the second year of synaptic blooming, more of the brain's neuronal fibres become myelinated. With these fatty insulations, nerve impulses get faster. Together with more synapses, this makes the cognitive potential of the young brain very high indeed, and usually corresponds to a boost in speech and language. An explosion of word use can make the vocabulary quadruple within a comparatively short time. At the same time, more complex and social skills develop, particularly those associated with the prefrontal cortex. The child learns to appreciate their own emotions: they will recognize themselves in a mirror, use their own name and have a sense of possession. They will also be better able to understand the progress of time, and therefore the link between cause and effect. This means they can learn how events in the past can relate to the present. At the same time, the deeper emotional centres of the brain – their limbic system – will be maturing, helping the child to make emotional connections with their parents and other children.

The peak in synaptic blooming in the brain means that a child will be affected by memories and learning far more in the first three years of its life than they ever will again, even though very few of us retain clear memories of this early childhood stage when adult.

By the age of three, motor skills have developed sufficiently to allow a child to manipulate objects. At this age, the brain is constantly learning new things.

The Brain Through Life

The Brain in Later Childhood

More brain development is focused on the prefrontal cortex as the child grows older. This part of the brain, covering the cerebral cortex across the frontal lobes of the cerebral hemispheres, is involved in controlling the highest cognitive functions of all. Synaptic connections continue to be made and pruned, while more myelination keeps the circuits running efficiently. Children are able to think more easily in abstract ways: they use their imaginations more and invent stories. And it is during this time that they will forge interests that may occupy them for the rest of their lives. At the same time, fine motor skills are being refined as synaptic connections are being made and broken in the cerebellum at the back of the brain.

Gender-specific differences, influenced by sex hormones, also tend to become more marked. These differences reach their peak with puberty, when new surges of hormones will have an important impact on the capabilities of the brain as children enter adolescence.

By age six, the brain is able to learn complex skills such as playing the piano. The composer Wolfgang Amadeus Mozart (1756–1791) was already composing pieces and performing in public at this age!

As children grow older, their imaginations develop and they are able to invent stories and create artworks.

The Adult Brain

Adult life takes us from adolescence to the grave, and leads us through times when our accumulated experiences, and memories of them, may help our brain achieve is greatest accomplishments.

As we emerge from puberty and pass through the unsettling time of adolescence, our brains have been fashioned ready for the long adult life ahead of us, but it is not until our early twenties that the final circuitry of the cerebral cortex is put in place. By then, our brains are controlling our behaviour more rationally. We are prepared for the responsibilities of being an adult, possibly even raising a family of our own.

Ageing brings with it more memories and accumulated experience, all of which affect our brains in one way or another. At the same time, the emotional charge that comes with events may lead us through periods of joy and despair. These influence the pleasure centres of the brain, or result in periods of stress. They are associated with surges of chemical signals – the neurotransmitters and hormones – that may affect how we think and what we do. When we reach the age of 40, our physical brain starts to lose volume. This is possibly caused by the shrinking of grey matter, although exactly what is happening at the level of cells is uncertain. However, despite this, the brain continues to function in a remarkable way. And by the time we reach old age, it is still far more complex than the most powerful computer.

The Adolescent Brain

The prefrontal cortex at the surface of the front lobes is the most important part of the brain involved in rational decision-making. It ensures that the choices we make that guide our behaviour are planned well and are even the morally right thing to do. This area of the brain is only fully mature by the time we are in our early twenties, and when our period of adolescence is behind us.

Rising levels of sex hormones during puberty bring on rapid changes in our bodies and can make us prone to become moody.

Adolescence is the transition into adulthood. The cerebral cortex is still developing, as the synaptic pruning in the grey matter that started in early childhood nears its completion. At the same time, the circuits of white matter are being strengthened. Neurotransmitters involved in regulating mood, such as dopamine and serotonin, surge, and this produces swings in mood and behaviour. The human brain is also influenced by the rise in sex hormones that were associated with bringing the body into puberty. As well as triggering sexual behaviour, testosterone increases aggressiveness and risk-taking. And all this goes on

The Brain Through Life

while the prefrontal cortex is being finely tuned to make our behaviour more reasonable.

The chemical effects largely override the efforts of the prefrontal cortex, so adolescents find it more difficult to plan, organize and make good decisions. This is why, until adolescence has passed, parents might be described as the surrogates for the prefrontal cortex, attempting to restrain more impulsive aspects of adolescent behaviour. So why do these adolescent difficulties arise in the first place? Are they just side effects of raging hormones? It is possible that, at least in the prehistoric past, they served a practical purpose. By making family relationships so strained, perhaps they helped to drive fledgling adults away from the family nest, encouraging them to breed with genetically unrelated strangers.

Emerging into Adulthood

Early adulthood can be the most productive time of our lives. Now that the cerebral cortex is fully developed, cognitive ability has reached its peak performance. This corresponds to a time when adult sexual drive is highest too. This is a good time for raising a family, even though the brain is still relatively inexperienced when it comes to memories and learning. However, with peak cognitive ability come peak learning skills, so we learn fast.

Physical Changes in Adulthood

After the age of 40 years, the brain shrinks in volume, losing about 5 per cent with every passing decade. The exact cause of this is uncertain, but it seems to be associated with neurons in the grey matter gradually dying and loss of the fatty myelin sheaths that insulates their fibres. However, it is also possible that growth of new synapses may happen to help compensate for this. Some parts of the brain are more affected than others: the prefrontal cortex changes the most, while the occipital cortex (involved in vision, at the back of the brain) changes the least. Other areas of the brain that suffer moderate shrinkage are the temporal lobes (involved in speech), the hippocampus (for memory) and the cerebellum (for coordinating complex movements). Some studies have suggested that the rate of volume shrinkage increases over the age of 70.

Like other areas of the body, the blood circulation of the brain becomes less efficient as we grow older. Blood flow through the brain's vessels decreases, which means that it supplies less nourishment and oxygen. This, in itself, reduces the performance of the neurons, especially since their activities make their energy demands so high.

By our mid-20s, our brains have fully matured, leaving us well prepared for the complex task of starting a family of our own.

Memory Loss and Gain

As we saw in chapter 8, memories are formed as physical connections between neurons. As neurons are stimulated in the short term, their electrical activities and pulses of chemical neurotransmitters persist fleetingly as traces, but are then fixed into more permanent synapses in the long term. This transition happens in a part of the brain within each cerebral hemisphere, the hippocampus, before being stored, over weeks or years, in the cerebral cortex. As we grow older, we live through more experiences and accumulate more memories, but the capacity for retaining them changes. The hippocampus is one of the parts that shrinks, so overall memory-processing declines. But there are subtle differences in the kinds of memory are involved.

Two kinds of memory change with age during adulthood: episodic memory and semantic memory. Episodic memory accounts for much of the memory decline. This covers memories that are tagged to episodes linked to where and when they were formed. An episodic memory could be about your first day abroad, or about a work meeting, or an interview for a new job. Some of these episodes will carry more emotional charge. This depends upon how the brain's limbic system was involved in processing the information (see chapter 9). The emotional memories are the ones that are more likely to be remembered, such as your first date with a partner, but, compared with semantic memories, episodic memories are more likely to wane as we get older.

Semantic memories are more about meaning and knowledge of facts. Remembering that hydrogen in the first element in the periodic table, or that Wellington is the capital of New Zealand are examples of semantic memories. And, remarkably, these memories seem to increase gradually with age. The reasons for this are not clear. Only in the very elderly are there reductions in semantic memory.

Chemical Changes in the Adult Brain

To a large extent, the activities of the brain are determined by levels of chemical neurotransmitters. Some of these excite neurons while others inhibit them, and the balance between them affects what the brain does (see chapter 6). And when neurotransmitters are active, synapses respond accordingly by getting reinforced and disintegrating. Two neurotransmitters, dopamine and serotonin, seem to be especially significant in the ageing brain. (These are the same chemicals that surge during adolescence: see above.) Dopamine levels drop by about 10 per cent with each decade after early adulthood. It is also possible that dopamine receptors fall too, so the brain is less receptive to the stimulating effects of this neurotransmitter. The overall effect is that the brain becomes weaker at processing information: cognitive and motor skills decline. A drop in serotonin may additionally affect the ability of neurons to form new synaptic connections. On top of this, changing levels of sex hormones may cause decline in performance. When a woman passes

An older university teacher is likely to have a wider semantic memory than his students, remembering more facts and figures.

The Brain Through Life

Stem cells in the brain, only recently identified, are used to regenerate new brain cells: both neurons and supporting glial cells. They are located in the hippocampus and ventricles of the brain and must migrate to where they are needed when they differentiate. However, only 50 per cent of them do so. The rest die.

through her menopause, her ovaries produce smaller amounts of the sex hormones oestrogen and progesterone. It is possible that a fall in oestrogen levels may reduce the brain's receptivity to dopamine still further.

What Causes Changes in the Ageing Brain?

The cause of brain shrinkage in old age is not easy to identify, and is likely to be the result of a number of factors coming together. A decline in neurotransmitters may lead to loss of synapses as they are used less and less, but what is making the neurotransmitters fall? In fact, the brain is just one organ that performs less well as we grow older. And, like other cells all around the body, neurons suffer from specific ageing effects.

One of the most significant changes that happen in dividing cells is that their chromosomes shorten with each successive division. Chromosomes are tipped with special caps, called telomeres, which protect the genetic information from the effects of this erosion. But there is a limit, and once cells have passed through a succession of divisions, the telomeres have worn down. At the same time, each division accumulates replication errors. All this can introduce genetic defects that make cell performance decline. Until the recent discovery of dividing brain stem cells (see chapter 4), brain neurons were thought not to be replaced. But the brain's glial cells also routinely divide, so are at risk from eroding telomeres. And, with these genetic effects, comes a decline in overall cell performance: cells are less able to process energy from food, respond to cues, absorb nourishment, and so on. It is likely that physical decline in brain cells is just another expression of the inevitable decline that happens everywhere else in the body.

Dementia

As brain performance declines in old age, this can impair day-to-day activities, and the brain is said to be suffering from dementia. This is associated with significant memory loss and can weaken thinking skills. It becomes more difficult to think quickly and clearly, and sufferers have difficulty making decisions. Dementia can also affect how we socialize with other people: sufferers may have trouble appreciating the moods and emotions of others. These are the symptoms associated with decline in particular parts of the brain involved in reasoning, memories and emotion, such as the prefrontal cortex, hippocampus, and other parts of the brain's limbic system. The incidence of dementia rises with age: to about 20 per cent at the age of 80, and to 40 per cent at 90.

More than half of dementia cases are caused by a particular disease: Alzheimer's (see below). The rest have vascular dementia, where dementia is

Alzheimer's Disease

In 1901, German psychiatrist Alois Alzheimer was shown the case of a 50-year-old woman called Auguste Deter who was suffering from early signs of dementia. When she died five years later, Alzheimer described her history and introduced the world to the disease that would carry his name. Alzheimer's disease is caused by distinct changes in brain tissue, although the diagnosis can only be properly confirmed post mortem: the neurons develop tangles of protein, while dark spots, called plaques, arise between the cells. It starts in the cerebral cortex of the temporal lobes before affecting the hippocampus, the brain's memory-storing centre. It then spreads through the rest of the cerebral cortex, affecting the occipital lobes last. Vision-processing is therefore the last to be affected, meaning that artists may retain their skills in the latter stages. But in the general progress of the disease, the entire cerebral cortex shrinks markedly.

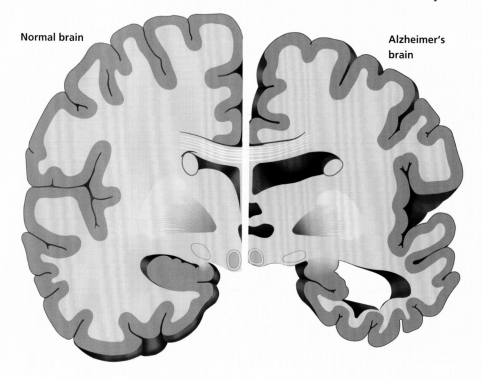

Normal brain

Alzheimer's brain

The Brain Through Life

caused by problems with blood circulation, although there may be overlap in cause and effect between the two kinds. There is some evidence that the risk of developing dementia is partly genetic, but environmental effects undoubtedly have a big impact too, with high blood pressure, diabetes and high cholesterol playing their part.

Staying Enriched

Ageing is, of course, inevitable, but painful dementia perhaps is not. We can at least take steps that help our brains stay working and active right to the very ends of our lives. Diet and exercise, unsurprisingly, play an important part. Foods that are high in energy, such as carbohydrates and fats, may increase the risk of decline, while antioxidants decrease it. These are molecules that mop up chemicals called free radicals that can otherwise interfere with the metabolic reactions inside cells, including the DNA that is so vital in controlling what cells do. In this respect, anything that keeps our entire body healthy will keep the brain working properly too. But with the brain, something else is needed too.

No other part of the body responds to stimuli like the brain. From the very first cluster of neurons in the microscopic embryo, to its older, wiser, adult form, the brain is plastic: life's experiences make its neurons connect and reconnect as they store up memories, and the richer the stimulation, the more they adapt and store. Research has shown that experimental rats suffering from Alzheimer's disease not only had improved cognitive skills when their cages were enriched with playthings, but that their brains showed physical improvements too. Similar studies indicate that direct electrical stimulation of the brains of human sufferers also improved thinking skills and mood. Just as enriched children end up with bigger brains, so mental exercise throughout life is the way to keep our brains plump and healthy. After all, the brain is the most complex thing in the known universe, so it's little wonder that looking after it is so demanding.

It is important for us to stimulate our brains throughout life. As we grow, this helps the brain to develop to its full potential. As we age, it helps the brain to maintain a high level of function.

Index

emotions 17, 20, 22, 46, 64, 65, 80, 109, 113, 115, 122–30, 144, 146, 178
  chemistry of 122
  emotional brain 131–6
  expressing 127
  and hormones 157
  and memory 120, 132, 133, 134, 182
  origin and purpose of 133
  pleasure 124–5
empathy 128–9, 136, 140, 142, 144, 152
encephalization quotient (EQ) 144–5
endocrine system 66, 67, 74, 76
endorphins 32
energy 10–11, 26, 46, 48, 70, 80, 88, 93, 181, 183, 185
environment
  and development 114–15, 177
  and emotions 130
enzymes 31
ependymal cells 47, 51
epidural space 52
epilepsy 28, 40, 98, 106
episodic memory 182
epithalamus 65, 67, 85
euphoria 90, 125, 160
evolution 8, 13, 14, 61, 64, 131
  of the brain 138–43, 140–1
executive functions 108, 146
exercise 72, 80–1, 87, 185
experience, learning from 113, 134
expiratory centre 83
extroversion 130
eyes 22, 38, 40, 66, 85, 92, 93, 94, 96, 99, 119

follicle-stimulating hormone (FSH) 155
food 24, 46, 48, 70, 83, 95, 113, 123, 133, 185
forebrain 57, 58, 64, 65, 167, 168, 169
fornix 132
fossils 138, 140
free radicals 185
frequencies, brain wave 104
frontal lobe 107, 108, 110, 140, 169, 179
fungi 25

Picture Credits

Shutterstock: 8 g-stockstudio, 10 Edgieus, 17l arek_malang, 17r lightpoet, 18l Gregory Johnston, 18r Mateusz Atroszko, 20, 155tl Tefi, 21, 107 Blamb, 22 Victoria Shapiro, 23br BlueRingMedi, 24 tl Sebastian Kaulitzk, 24tc Aldona Griskeviciene, 26 Svoboda Pavel, 28 Yakobchuk Viacheslav, 29 Anatomy Insider, 32l MaLija, 32r Maridav, 34 Alex Mit, 39tr Alila Medical Media, 40 MidoSemsem, 44 andregric, 46 Chuong Vu, 62 Raluephoto, 63 Kateryna Kon, 64 Nicolas Primola, 66 Designua, 68 Sebastian Kaulitzki, 70 Dean Drobot, 74 sirtravelalot, 79tl stihii, 79tr Sakurra, 80 Rihardzz, 82tl Designua, 82tr, 83br Vasilisa Tsoy, 83bl Mari-Leaf, 84 bbernard, 88 Magcom, 92 g-stockstudio, 95, 96, 97 Alila Medical Media, 98 ellepigrafica, 104 Ponkrit, 107 miha de, 110 Sebastian Kaulitzk, 114 Sokolova Maryna, 116 Irina Kozorog, 117 NatalieJean, 122 SingKham, 124 Linnas, 126 Belish, 128tr Wayhome Studio, 128tc Sfio Cracho, 128tr Mix and Match Studio, 128bl Rachata Teyparsit, 128bc ArtFamily, 128br Aleksandra Kovac, 131 worldswildlifewonders, 133 Eric Isselee, 135, 144, 174 Monkey Business Images, 136 Rawpixel.com, 138 blvdone, 139l Magdanatka, 139r Zzvet, 140 Nicolas Primola, 141 top row l ivanpavlisko, r witoon214, 141 second row r, third row c, r Puwadol Jaturawutthichai, 143 Gorodenkoff, 146 Human Brain Anatomy, 148 Chompoo Suriyo, 154 Oltre Lo Specchio, 156 Lightspring, 158 TijanaM, 159 StockLite, 160 Pongsathon Ladasuwankul, 161 Luke Shelley, 163 Cam Craig Creative, 164 bodom, 166 Anneka, 167 Kateryna Kon, 169 HelloRF Zcoo, 171 GagliardiImages, 172 Steve Allen, 173 Timof, 175 Vecton, 176 Gorynvd, 178 MIA Studio, 179t Morphart Creation, 181 Rido, 182 sirtravelalot, 184 ilusmedical, 185 VGstockstudio

Dreamstime: 41 Alila07, 155 tr Bluezace

Other: 35, 56 Wellcome images, 54 Manu5/Creative Commons, 106, 177 National Institutes of Health, 108 Mark Dow, Research Assistant Brain Development Lab, University of Oregon, 112 Valerie/Creative Commons, 141 top row cr Nachosan/Creative Commons, 141 second row l Bjørn Christian Tørrissen/Creative Commons, cl Guérin Nicolas/Creative Commons, cr Daderot/Creative Commons, third row l Ji-Elle/Creative Commons, bottom row l, c Durova/Creative Commons, r Luna04/Creative Commons, 183 California Institute of Regenerative Medicine

Illustration Credit

Daniel Limon (Beehive Illustration)

Every effort has been made to contact copyright holders. The publishers will be pleased to make good any omissions or rectify any mistakes brought to their attention at the earliest opportunity.

Publishing Director *Trevor Davies*
Production Controller *Grace O'Byrne*

For Tall Tree Ltd
Editors *Rob Colson and Jon Richards*
Designers *Gary Hyde and Ben Ruocco*

Acknowledgements